HUMANESIS

posthumanities

Cary Wolfe, Series Editor

(continued on page 211)

HUMANESIS

SOUND AND TECHNOLOGICAL POSTHUMANISM

DAVID CECCHETTO

posthumanities **26**

University of Minnesota Press
Minneapolis
London

Portions of the Introduction and chapter 5 were previously published as "Deconstructing Affect: Posthumanism and Mark Hansen's Media Theory," *Theory, Culture, and Society* 28, no. 5 (2011): 3–33. Portions of chapter 2 were previously published in *Eiodola: William Brent and Ellen Moffat* (Victoria, B.C.: Open Space Arts Society, 2009). Portions of chapter 4 were previously published as "Melancholy and the Territory of Digital Performance," in *Collision: Interarts Practice and Research,* ed. David Cecchetto, Nancy Cuthbert, Julie Lassonde, and Dylan Robinson, 77–90 (Newcastle upon Tyne, U.K.: Cambridge Scholars, 2008); published with the permission of Cambridge Scholars Publishing. A different version of chapter 6 was previously published as "Sounding the Hyperlink: Skewed Remote Musical Performance and the Virtual Subject," *Mosaic* 42, no. 1 (2009): 1–18.

Published by the University of Minnesota Press
111 Third Avenue South, Suite 290
Minneapolis, MN 55401-2520
http://www.upress.umn.edu

LIBRARY OF CONGRESS CATALOGING-IN-PUBLICATION DATA
Cecchetto, David.
Humanesis : sound and technological posthumanism / David Cecchetto.
(Posthumanities ; 26)
Includes bibliographical references and index.
ISBN 978-0-8166-7644-6 (hardback)
ISBN 978-0-8166-7963-8 (pb)
1. Technology—Philosophy. 2. Technology—Social aspects.
3. Humanism—History—20th century. I. Title.
T14.C43 2013
303.48'3—dc23   2013001447

Printed in the United States of America on acid-free paper

The University of Minnesota is an equal-opportunity educator and employer.

20 19 18 17 16 15 14 13   10 9 8 7 6 5 4 3 2 1

CONTENTS

ACKNOWLEDGMENTS

I am extremely grateful to Stephen Ross, whose efforts and expertise in guiding me through the process of writing this book are matched only by his patience with my tendency to obfuscate doubly where I am asked to clarify; moreover, Stephen has acted as both an advocate and a mentor, and I set him as my standard when working with my own students. I would also like to thank Steven Gibson, particularly for sticking his neck out to make sure that I received the resources I needed for the creative projects that subtend this work. Christopher Butterfield models a life of thinking differently and delightfully, which will always be an aspiration of mine. Many faculty and students in the University of Victoria's program in Cultural, Social, and Political Thought have also played an animating role in this book, which couldn't have come to be without them. I'm also grateful to Cary Wolfe and Doug Armato for their generous spirits and discerning minds.

I am indebted to many friends. Dylan Robinson frequently catalyzes new directions in my research and has been an arbiter of taste and a spirited interlocutor throughout my personal and academic life. Anita Girvan provided more support than one should ever ask of a friend—from editing to counseling to any number of well-timed jokes—and even provided temporary housing under a roof with her sparkling family (I'm afraid all I gave in return was several hundred feet of rolled paper!). Marc Couroux and eldritch Priest have helped to keep me a practicing nonmusician, for which I offer thanks only in the form of a promise to continue to contribute to our collective bellyaching.

Finally, my family: my parents instilled in me the (at times misplaced!) self-confidence necessary to complete a project of this scope, and my siblings (humorously) enforce the humility that keeps me pushing further. Sevina and Jasmine have suffered my philosophizing of pencils graciously and have taught me to attend to the complexities of life with a joyfulness I never thought available to me; our time together has been the richest and most wonderful part of my life. And of course, to Anna, née Gertrude, who has been a muse throughout this project in ways she'll never know. I only hope this book has a fraction of the depth of her own work; I know that it pales in this respect next to my love and admiration for her.

INTRODUCTION

POSTHUMANISM(S)

> To hear past the historical insignificance of sounds, we need to hear more than their sonic or phonic content.
>
> DOUGLAS KAHN, *NOISE, WATER, MEAT*

It is in the character of sound to be semiotically parasitic, to take on—and usually intensify—the systems of meaning to which it attaches. High-fidelity audio accompanying a video, for example, tends to produce the impression of higher-definition visuals, while the reverse is not the case.[1] Sound is a kind of *amplifier,* but an invisible one, a contamination that produces something different precisely by reproducing the same.

A book about sound is never just a book about sound, then, but is rather a stridulation of sound rubbing up against another set of concerns, in this case, technological posthumanism. This book palpates three dominant strains of the latter, examining each as a contingent narration of human–technology coupling. To this end, the book adopts the critical *strategy* of seeking after the assumptions and biases that underwrite each approach, while also taking up the *tactic* of playing each narration out in a topos of sound, where we might listen more particularly to their politics. The gambit of this book *as a book,* then, is that the coaction of these strategic and tactical approaches might produce fresh purchase on a discourse that continues to proliferate, not least because the daily practices to which it obtains become increasingly difficult to even provisionally extricate from contemporary technologies. That is, *listening*—in the full sense—to technological posthumanism will not only offer new insight into what has been said on the topic but will also push the conversation in a direction that is crucial to (and, to date, largely missing from) the broader posthumanist project of decentering the human.

We can more specifically understand the challenge that aurality presents in this context by further considering three of its aspects, each of which points to the way that sound is mobilized in this book. First, sound is differential: as Aden Evens points out, "to hear is to experience air pressure changing. . . . One does not hear air pressure, but one hears it change over time [such that] to hear a pitch that does not change is to hear as constant something that is nothing but change. To hear is to hear difference."[2] The point, for Evens, is that the physics of hearing meshes perfectly with a Deleuzian language of becoming, but the inverse should also be noted, namely, that sound can only be understood as a physics insofar as it is a physics under erasure, as a physics that always performs itself and its own impossibility simultaneously. In this way, sound supplements (or perhaps tropes) the famously vexing horizontal duality of light (as both particle and wave) with a vertical duality that exists as both material (though exclusively in wave form) and immaterial. In short, sound as such calls us to think of it as a particular object that has no substance, as a kind of ideal object that nonetheless has real material effects (i.e., literal sounds).

This, then, points to the difference between attending to sound in its own right and listening to sounds: despite its undeniable material effects, *sound itself* resists being placed within a visual ontology. Indeed, sound resists being placed at all and is in this sense as much relational as it is differential (which is the second feature I wish to highlight). Think, for example, of a light panning across a stage from right to left; if we stop the light 80 percent of the way through the pan, we can point to the exact place where the particles of light literally impact the stage. Now imagine the same scenario with sound; with contemporary technologies, we can almost as easily pinpoint where the sound "is" on stage (i.e., where we hear it coming from), but there is in fact nothing there: the placement of sound that results from an 80 percent pan is in fact produced by a relative difference in intensity between the two polarized sound-emitting loudspeakers. In short, the sound is emphatically *not* where it sounds like it is. Indeed, the added twist is that it also isn't where it *appears* to be (i.e., coming from the loudspeakers) because it only comes to be at all through the differential act of hearing, which is the very act that would place it where it isn't.[3]

Finally, sound is also multiplicitous in that—in Deleuze and Guattari's terms—any given sound consists of $n - 1$ sounds. Thus, for example, it is in the nature of frequencies to consist in an infinite number of higher frequencies, with each successive frequency a determinable multiple of the basic frequency of the sound; it is these overtones (or "partials") that result in the same fundamental pitch sounding different when played on a violin or a piano, for example, or that allow us to differentiate between two voices singing the same pitch. That is, the relative intensities of the invariant series of overtones produce the timbre of a sound, sometimes called its "color." The reason why this is multiplicitous rather than simply multiple is that the infinite nature of this sequence—coupled with the environmental contingencies that are constitutive of any wave behavior— means that the original sound can never be perfectly reverse engineered from its constituent frequencies. In short, the original sound exists as a sound among other sounds rather than as a unity of its various overtones. The correlate to this is, of course, that sounds are also radically singular.

Taken together, these elements of sound mean that it's distinctly possible that an experience of aurality proper is precluded precisely via its transformation into an object of knowledge. This, of course, is familiar to us as the problem of *any* relation between experience and knowledge. The question, though, is whether the operational difference that is a condition of hearing—where to hear is literally to experience as constant something that is nothing but difference—might be mobilized productively in the context of technological posthumanism. Rather than going in search of the truth of sound, then, this book asks after the diffraction patterns that arise when specific aural experiences are rubbed against specific narrations of human–technology coupling. This is the reason that sound is approached obliquely in this text: though the preceding characterizations of sound may seem simple enough, fully embracing their impact—and sustaining this embrace—requires a method that takes these declarative statements about sound seriously without treating them as true per se, that is, without *knowing* sound. This is not an easy task: on one hand, we have been routinely lamenting the dominance of the visual as the privileged sensory and aesthetic mode of the "human" of humanism since at least Sartre's work on the Look in *Being and Nothingness*, which notes that the "core fantasy of humanism's trope of vision is to think that perceptual space is organized around and for the looking subject";[4] on the other hand, it remains the case that we haven't yet learned how to theorize the force of sound in its opposition to this modality, as an "undissolvable residuum" that hollows itself.[5]

IT IS PRECISELY the forceful quality of sound that makes it an agent of modulation that can help to amplify certain elements of narratives of human–technology coupling, making them audible. This is particularly crucial today, when the signs of technological posthumanism have become so ubiquitous that most of us, on most occasions, have ceased to take notice. Indeed, our navigations through diverse realities and our negotiations with aggressive technological couplings are no longer even really remarkable: e-mail and video chats, certainly, but also genetically modified foods, text messaging, online classes, virtual exercise routines, and complex relational databases are all part of our daily lives. As a result, our perceptive apparatuses are constantly tuned to diverse and often contradictory frequencies, but in a way that tends to play out any incongruities as part of what and who we are as humans. Indeed, many of us are more likely to feel the absence of these technologies than their presence: for example, I myself have frequently felt a visceral frustration at not having Internet access while driving; a question pops into my head and—before becoming aware of my physical situation—I can feel my body reaching for the Internet to answer it. What feels incongruent is not having the Internet available, as it forces me back into the confines of a body that I no longer identify as being the sum total of my self. Moreover, even as I feel this bodily recontainment as a reductive violence played out on my consciousness, I nonetheless continue to hurtle along the highway at a speed that insists that I, too, am propelled, just as much as my automobile.

Two things are revealed in this scenario, which is a classic example of how contemporary technologies disrupt the long-held assumption that "our bodies are isomorphic or at least 'proper' with ourselves":[6] first, that to the extent that our subjectivity pertains to our actions in the world, we are compelled to think of ourselves through the lens of technology; second, that this has always been the case. Of course, if the second point is true, then we can safely say that the first point is not a new one. However, one feature that separates our current mixed reality—the mixed reality of digital and analog technologies—from that of the past is precisely the fact that we are able to marvel at our own ability to navigate so seamlessly from one realm to another. As Mark Hansen notes, the question of how we can accomplish this feat so fluidly *"did not need to be posed* so long as perceptual experience (with only atypical exceptions) remained within a single experiential frame—so long, that is, as experience typically occurred within a single perceptual world as a coupling to a single form of extension or homogeneous outside."[7]

Thus, with the dramatically increasing prevalence of digital technologies in contemporary Western culture, the end of the twentieth century witnessed an important shift in the poststructuralist project of the deconstruction of the subject. The otherness of the Other that is constitutive of the modern subject, for example, is now understood to be predicated on *technologies* of reflection and language. Technology has entered into the discourse at the ground level. Though this is in itself not particularly novel—the *linguistic turn* might be understood precisely as the recognition that discourse is always-already in some sense technological—the evolved relation with technology that characterizes contemporary culture strikingly reorients the terms in question. This new cultural inclination is what is captured under Stephen Johnson's nomination of our present historical moment as *interface culture,* a term he "wields to embrace not only the ubiquity of computers and electronic devices but also the way in which interface has come to function as a kind of trope or cultural organizing principle."[8] More radically, Lev Manovich has positioned the (relational) database as a symbolic form that has replaced, in contemporary Western culture, the privileging of conventional linear narrative that operated in the age of cinema. In the logic of the database, individual items are collected in such a way that "every item has the same significance as any other,"[9] so that the signification processes of traditional causal sequences are severely troubled.

What has changed, then, is not simply the technologies with which we interact but our conception of technology itself. As numerous theorists from diverse disciplines have argued, technologies today can no longer be adequately thought through the lens of "extension" but must instead be understood as profoundly implicated in our being. That is, technologies are not tools that we use, nor objects in relation to which we are servomechanisms, but are rather pathways through a relational ontology (which may be another way of saying that technologies are also all those things that they are not). This is the sense in which the subject is thought as a technology in this study, where the latter indicates (like Agamben's "apparatus") "literally anything that has in some way the capacity to capture, orient, determine, intercept, model, control, or secure the gestures, behaviors, opinions, or discourses of living beings."[10] In short, rather than being characterized in terms of the subject's relation to the social, today's subject is first thought in relation to technology. As a result, N. Katherine Hayles seems to have been correct in her prediction that twenty-first-century debates "are likely to center not so much on the tension between the liberal humanist tradition and the posthuman, but on different versions of the posthuman as they continue to evolve in conjunction with intelligent machines."[11]

Indeed, this recursive feedback loop is actually contained in McLuhan's well-known tetradic reading of technology, which describes individual technologies (including "software" technologies such as languages and ideas) as nodal points in a field comprised of enhancement, obsolescence, retrieval, and reversal potential.[12] E-mail, for example, might be said to enhance one-to-many communication, but pushed to its extreme, this enhancement flips into its opposite, which we know by the name of "spam." Similarly, e-mail can be said to obsolesce letter writing in favor of the interactivity and informality that had previously been dominant in oral communication, which it retrieves in a new form (importantly, letter writing nonetheless persists precisely through its obsolescence, taking on a new quality of almost unavoidable intimacy in this new context). For McLuhan, these four "laws of media" collectively stem from the premise that "each of man's artefacts is in fact a kind of word, a metaphor that translates experience from one form into another,"[13] so that artifacts of both hardware (i.e., a table, or a stick) and software (i.e., an idea, or a preposition) nature are understood by McLuhan as extensions of a human body (which includes that body's mind). However, the simultaneity of each vector—that is, the fact that every enhancement carries with it a *simultaneous* reversal potential, obsolescence, and retrieval—in combination with the present ubiquity of intelligent machines has ultimately decentered this human body, specifically by suggesting that the body itself, as well as subjectivity, is invested by technology.

If subjectivity can be understood as a particular and fluid combination of McLuhan's four intensities, then questions of identity are replaced by (or at least supplemented with) questions of procedure. As a result, the discourse of technological posthumanism—inasmuch as the term can capture what is a diverse and often contradictory assemblage of thinkers— is not a project of redefining subjectivity, humanity, or the individual so much as it is one of considering the conditions, exclusions, and performativities of these terms. Thus my research treats a discursive wrangling around the subject—or "selfhood"[14]—as the first meaning of "technological posthumanism," an approach that emphasizes "the boundary-making practices by which the 'human' and its others are differentially delineated and defined."[15]

Whatever (technological) posthumanism may or may not suggest, it is clear that the term registers a coimplication of technology and the subject, what Mark Poster has described (using the term *humachine*) as an "intimate mixing of human and machine that constitutes an interface outside of the subject/object binary."[16] In thus registering the term first as a nexus of discursive relations, I take my cue from Herbrecter and Callus, who argue that

> *posthumanism,* as the name of a discourse, suggests an *episteme* which comes "after" humanism ("post-humanism") or even after the human itself ("post-human-ism"). Implicit in both these articulations is a sense of the supplanting operations wrought by time, and of the obsolescence in question affecting not simply humanism as displaced *episteme* but also, more radically, the notion and nature of the human as fact and idea.[17]

Indeed, Neil Badmington goes a step further in arguing that—in light of critical theory since at least Lacan—"humanism never manages to constitute itself [but instead] forever rewrites itself as posthumanism."[18] As a result, although the term *technological posthumanism* does not indicate something fixed or concrete, it nonetheless captures the relation of humans and technology in a way that allows it to act as the basis—the starting point—for further investigations in other areas. Considered in the context of the ever-increasing rate of proliferation of new technologies, this meaning may well chart the path of humanity's futurity, declaiming the terms of registration for new ideas and new ways of thinking the limits of possibility.

All the preceding points to what is indicated by the neologism that titles this book, *humanesis*: the suffix *-esis* can indicate both a state and a process, and posthumanism in the broadest sense might be initially defined as a recognition that the static term *human* has entered into discourse, where it flows, mutates, amplifies, exchanges, and propagates according to the various and often paradoxical logics of language. What I am calling *technological* posthumanism, then, considers more specifically how the subtending discursive vectors of technology modulate these movements, altering their affordances. In this way, considering "the human" discursively is not a reduction but rather an avowal, and one that emphasizes (as I demonstrate throughout this book) deconstruction as a performative theory of media and meaning in the broadest possible sense.

As I will argue throughout this text, then, what is at stake in each of the competing versions of technological posthumanism that are considered herein is the perspective from which such changes are registered. That is, each perspective is supported by a disavowed system of values—by a specific set of assumptions—so that what must ultimately be thought are the ways in which these values are sublimated and normalized by each perspective and what is at stake in each competing strain's emergence as a potentially dominant cultural logic.

IF SUBJECTIVITY is to be approached in terms of technology, then it follows that media art—art involving contemporary (usually digital) technologies—offers significant insight into the specific constitution of contemporary subject positions. Hayles supports this view, remarking,

> If art not only teaches us to understand our experiences in new ways but actually changes experience itself, [new media] artworks engage us in ways that make vividly real the emergence of ideas of the body and experiences of embodiment from our interactions with increasingly information-rich environments. They teach us what it means to be posthuman in the best sense.[19]

Indeed, the existence of at least some minimal relation between art and subjectivity has long been established so that Hayles's point here is basically that the tradition of studying the subject (or individual) in light of its contemporary artistic and literary output has been maintained through the transition to digital culture. What is strange, though, is that even though early characterizations of *electric culture* (the embryo of digital culture) regularly operated through critiques of the privilege afforded to vision in contemporary culture[20]—and even though the worldviews that have sprung from refocusing on haptic and sonic perceptual apparatuses have been mobilized against this hegemony—there remains a relative dearth of study that focuses specifically on the role (and constitution) of sound in media art's performances of the subject (this despite the relatively recent emergence of sound studies as a field of inquiry unto itself). To understand the extent of this omission, consider that as recently as 2004, Johanna Drucker argued that "the idea that *visual* representation has the capacity to serve as a primary tool of knowledge production [was at that time] an almost foreign notion to most humanists."[21] If the digital humanities are currently only in the nascent stages of contesting the exclusion of the visual in its own right (i.e., in its extratextual mediality), the aural has been almost entirely neglected.[22] Thus, although each of the theorists who acts as a nodal point in this study has devoted significant effort to reading artworks through his or her theoretical model, the works are only rarely considered in terms of their sound. This exclusion implies, to my ears, that sound in some way threatens this discourse; this book goes in search of the politics and implications of this exclusion.

There are, of course, exceptions to this omission, theoretical noises (in the best sense) often sounded by thinkers with one ear tuned to undermining the hegemonic (disciplinary) politics of music. Christoph Cox, Paul Hegarty, Douglas Kahn, Brandon Labelle, Paul D. Miller (aka DJ Spooky), Charles Mudede, and Jonathan Sterne all variously exemplify this important work. However, when Mudede (for example) thinks the turntable as "a repurposed object [that is] robbed of its initial essence [but] is soon refilled by a new essence,"[23] what is being thought is not really specific to the *sound* of the turntable. That is, Mudede's turntable is foremost a cultural artifact rather than a medial one in that everything ascribed to the turntable could conceivably take place without the medium of sound (though perhaps not without music).[24] This difference points to a necessary clarification of my own project: my mobilization of sound in this study is not intended to intervene with musical discourse nor to claim a privileged status for sound. Instead, this book takes advantage of sound's pronounced mediality—in the sense that sound art "has nothing but mediations to show for itself"[25]—to trouble the assumptions that underwrite the theorists in question. After all, sound remains (rightly or not) the test of presence, even if Derrida has taught us that this is a test that will always fail.

THE BODY OF THIS BOOK is a critical discourse analysis of three strains of technological posthumanism that emphasizes the cultural–political stakes of each, the way that each instance of humanesis—the putting-into-discourse of the human—directs our understanding of human–technology coupling along different evaluative vectors. This emphasis is lent support by readings of contemporary media artworks that are intended to probe a specific (posthumanist) problematic raised with respect to each theorist and to avow the study's own unavoidable role in constructing the discourse that it describes. To this end, these artistic analyses are attentive to the medial specificities of the works they consider and give particular emphasis to the role that sound—as introduced at the beginning of this introduction—plays in their relational networks. Taken together, then, technology is at the center of every level of this project: as the forms of selfhood that are exemplified in each strain of technological posthumanism (i.e., technologies of subjectivity); as the material conditions and (often) the aesthetic ground of the media art practices considered; and as the mode in which these considerations are collected into a unified text. Pairing the centripetal force of theoretical close readings with the (imaginary but no less visceral) centrifugal flights of artistic practice, this project introduces a new, procedural understanding of the inductive theoretical knowledge that is already in play at the junction of technology, media art, and subjectivity. In short, this approach attends to the particularities of cultural production in their own right, emphasizing not only the ways that theory is supported by cultural practices but also the ways in which the latter tend to elude theoretical discourse. This understanding not only enhances the theoretical milieu within which this study operates but also transfigures it through a rejuvenated emphasis on the praxis of meaning-formation in its inductive capacity.[26] The theoretical content of this book is focused around three thinkers: Ollivier Dyens, N. Katherine Hayles, and Mark B. N. Hansen. Each of these three offers a different inflection to the study at hand, sending it in disparate directions: Dyens's *Metal and Flesh* intensifies Richard Dawkins's logic of "selfish" genetic reproduction in its cultural aspect, articulating the deterministic challenge to human agency implicitly issued by notions of scientific discourse constructed around measurability, repeatability, and falsifiability. Hayles has been a primary figure in the rise to discursive prominence of technological posthumanism, and the trajectory of her thought charts a genealogy of the future of the technological posthuman that registers the specificity of contemporary technologies in their own right. In this respect, Hayles's attempts to think beyond what she perceives to be the limits of deconstruction are read here for the contributions that they offer to the discourse, despite her project ultimately being tethered to the humanist values that she seeks to gain traction against. Finally, Hansen's advocacy for an extralinguistic understanding of embodiment and technics moves against deconstruction from the other side, articulating a fully present body that would precede linguistic ambivalence. Though Hansen's analyses give an important account of how an operational perspective can inform contemporary thinking about couplings of humans and machines, I argue that the conclusions that he draws from this perspective reinforce (rather than undermine) the tenets of Derridean deconstruction.

Similarly, the artworks considered in tandem with these theorists also each offer a unique perspective. The analysis of *Eidola*—an exhibition featuring works by William Brent and Ellen Moffat—teases out the exhibit's challenge to scientific (visual) logic from two sides, showing both that it is haunted by forces that are qualitatively different than those that it registers and that its very signs, pushed to the extreme, turn back on themselves to speak against the terms that conditioned them in the first place. Similarly, the analysis of Rafael Lozano-Hemmer's *The Trace* reads the telepresence of the piece as demonstrative of the type of complexly contingent agency that Hayles unpacks but also emphasizes (particularly when thought in tandem with Judith Butler's theory of melancholic subjectivity) the ethical ambivalence that obtains in the performative dimension of this situation.

Moreover, since *The Trace* is an exemplary piece of early telepresent art, studying it gives purchase on media art's points of departure from traditional disciplinary practices as well as the ontological assumptions that are built into these practices. Finally, the analysis of Skewed Remote Musical Performance (SRMP)—a collaborative practice between William Brent and me—exemplifies the rich entanglement that Hansen deems characteristic of our present historical moment but also reiterates the extent to which meaning construction remains, even in the context of ubiquitous media, conditioned by linguistic practice. In this context, if Hansen's thought points a digital finger back toward analog subjectivity, SRMP might be said to perform the inverse gesture, insisting on the impossibility of inscribing an originary status to either analog or digital realities.

Clearly, if the theoretical chapters of this book are aimed at thinkers whose influence in their field is—to varying degrees—self-evident, the same cannot be said of the artistic analyses found in the accompanying chapters. Although ample justification for each selection can be found—I believe—in the analyses themselves, a more general note about their selection may also be helpful: quite simply, these works were not chosen in an attempt to articulate a canon of new media art, nor because they are particularly "important" in a grand historical sense; instead, the artistic chapters of this book move in the doubled motion of lines of flight, on one hand deterritorializing the relatively linear narrative that otherwise characterizes the book, while on the other hand reterritorializing my own position in the study and thereby acknowledging my role—as a seemingly objective analyst—in constructing the facts that I seek to describe. With respect to the former movement, these works are each intended to remind the reader, in different ways, that this study does not aim to contain the discourse of technological posthumanism but rather to unpack an instance of it. Regarding the inverse trajectory, the artworks are selected in part because they are each at a different personal distance from my own artistic practice: I have a conventional objective relation with *The Trace,* a profoundly personal relation with SRMP (as its co-creator), and a less personal but still deeply informed relation with *Eidola* (as co-curator of the exhibition). By thus situating myself in multiple ways, I hope to acknowledge my own subject positions but also to open a space of indetermination within the broader narrative(s) of the book. That is, I have taken it to be a truism throughout this study that content is always context specific; in selecting works that access qualitatively different strata of knowledge (within myself), I hope that this will be all the more explicit. Put differently, these chapters support the conditions for second-order observation—or "observation of observations"[27]—that serve to more expressly situate them in the study, but their particularities suggest that the reverse is also true, namely, that the discourses of technological posthumanism discussed offer fresh perspectives on many of the salient concerns of media art. Thus the ambivalent relation of the artistic and theoretical chapters—themselves also internally ambivalent—ultimately also redoubles the emphasis on performativity that obtains throughout the book, thereby desublimating the "indissoluble mingling"[28] of aesthetics, technology, subjectivity, and ideology that each performs.

CHAPTER 1, "From Genes to Memes: Ollivier Dyens and the Scientific Posthumanism of Darwinian Evolution," reads the detachment of embodiment from biology described in the theoretical work of Ollivier Dyens.[29] In *Metal and Flesh*, Dyens updates Richard Dawkins's "selfish gene" argument to insist that the dominance of information in our current media-rich environment results in a human body that exceeds humanist notions of embodiment. In this context, Dyens frequently cites the cyborg as the posthuman body par excellence, claiming that the cyborg is a "living being whose identity, history, and presence are formulated by technology and defined by culture."[30] In this characterization, there is perhaps an inflection of advocacy in Dyens's writing, eliding with the transhumanist perspective that we are evolving rather than, for example, shifting toward—or always-already implicated in—technological posthumanism.[31] Indeed, the notion of evolution is central to Dyens's thought, which argues that our species's modus operandi—which he believes is the desire to survive and to reproduce—has shifted from a biological register to a cultural one, such that desire is now configured around seeking out "culturally fertile bodies."[32] What arises from this perspective, though, is a constitution of "life" that risks tautology: if life is defined, a priori, in terms of evolution, what is really being said when we chart evolving processes outside of the traditional domains of the living as constituting life? I argue that this looming tautology—of which Dyens is certainly aware—constitutes a primary driving force in *Metal and Flesh,* leading Dyens to the necessity of pointing to science as a purveyor of certain measurable truths, even as he retreats from these very claims.[33] Thus this chapter treats Dyens's understanding of life itself technologically, asking "what kind of regulatory apparatus it works in the service of"[34] and what constitutions of the posthuman it might foreclose. In short, this chapter asks what privilege there is to scientific knowledge and how this privilege is written into Dyens's technological posthumanism.

Ultimately, chapter 1 conducts a hauntology of the positivist definition of life that Dyens extrapolates from Dawkins. In chapter 2, "Dark Matters: An *Eidolic* Collision of Sound and Vision," the invisible forces that this haunting suggests are further explored by considering the mixed-media exhibition *Eidola* in relation to two related but distinct metaphors: ghosts and dark matter. From these tropes, the chapter argues that *Eidola* stages an encounter between disciplinary biases of sonic and visual art practices, accentuating how both are infused with a part of the other that they cannot avow. Showing that the sound of *Eidola* is a blind spot in its visual observation, the chapter argues that this operates equally to trouble both the exhibit's sonic and visual components, even as it asserts the relation between the two. Ultimately, then, sound intervenes in this reading as a dual identity that is simultaneously Other and interior to vision: whereas dark matter is visible only through its invisibility—through its gravitational pull on visibility itself—*Eidola* suggests that we might instead *hear* dark matter as a kind of sonic delirium that, rather than being a structuring principle, puts the lie to structure itself.

In a sense, both of the first two chapters revolve around a scientific means of constructing knowledge that has contributed to a contemporary understanding of "information," with Dyens intensifying this logic and *Eidola* troubling the positive substantial claims that it implies. Moving this line of inquiry in a new direction, chapter 3, "N. Katherine Hayles and Humanist Technological Posthumanism," engages the work of Hayles in its attempt to think beyond deconstruction. To this end, the chapter comprises close readings of key terms in Hayles's popular and well-respected "posthuman trilogy" of *Writing Machines, How We Became Posthuman,* and *My Mother Was a Computer,* giving particular attention to her renderings of intermediation, distributed cognition, and embodiment as well as to the ways that her conception of materiality as an "evolving property created through dynamic interactions"[35] evolves over the course of the texts. Ultimately, the chapter argues that Hayles's enormously influential construction of technological posthumanism is best read as a recombinant humanism that more readily connects (for better and worse) to an established system of human values than it does to the operations of the media that are its subject. In this, Hayles aligns with her fellow literary critic Fredric Jameson's belief that "narrative itself is the inevitable means by which we attempt to make sense of the Real of history: we don't have to narrate the way we do, but we do have to narrate."[36]

Chapter 4, "The Trace: Melancholy and Posthuman Ethics," mobilizes Rafael Lozano-Hemmer's participatory telepresent installation The Trace in relation to the call for an embodied understanding of information pronounced in Hayles's technological posthumanism. In particular, this problematic is taken up by reading The Trace with and against Judith Butler's account of melancholic subjectivity, specifically as it is articulated in Antigone's Claim.[37] Through this lens, I argue that the subjectivity performed in The Trace unilaterally reduces the participants' modes of relating to one another. However, rather than either authoring a dematerialized body or evincing a priority of embodiment, this reduction allows the piece to function as a critique of the unilateral narratives that it performs and also of the symbolic form of relation itself as it obtains in Butler. As a result, The Trace exists in a tension with both vectors of (de)materialization that, ultimately, positions their relation as the terrain of its posthumanist ethics. In turn, this poses a significant challenge to Hayles's (foundational) attachment to the possibility of nonhegemonic meaning, reemphasizing the containment of her thought in a language of value.

Chapter 5, "From Affect to Affectivity: Mark B. N. Hansen's Organismic Posthumanism," examines a construction of technological posthumanism that, in a sense, attempts to move beyond such containment by prioritizing an operational perspective. Whereas Hayles registers the complex intermediating feedback loops that compose the relation between bodies and the world of technology, Hansen sidesteps this problematic in favor of a notion of "primary subjectivity" in which the subject is constituted through affective spatiality (or "affectivity"). In this context, the chapter investigates Hansen's attempt to give a robust account of technology in its extralinguistic dimension by evincing an "'originary' coupling of the human and the technical" that grounds experience as such and that "can only be known through its effects."[38] Ultimately, the chapter finds that Hansen's perspective remains haunted by the representational logic that it moves against. However, this observation does not repudiate Hansen's argument as such but rather rejects one of its central underlying implications: that the extradiscursive materiality of technology might be accessed, linguistically, without biasing it in a way that is foreign to this materiality. To this end, the chapter articulates Hansen's argument for an affective topology of the senses, corroborating the increased importance of digital technologies in this perspective through a brief comparison of Robert Lazzarini's *skulls* (as read by Hansen) and my own piece *Sound*. From this comparison, I ultimately argue that what is accomplished by Hansen's putting-into-discourse of *technesis* is, paradoxically, a restaging (and perhaps even a heightening) of the constitutive ambivalence of deconstruction that he seeks to undermine.

If the organismic posthumanism of chapter 5 performs an intensification of the paradoxical (deconstructive) causality that it disavows, chapter 6, "Skewed Remote Musical Performance: Sounding Deconstruction," discusses an art practice that makes this performance explicit. To this end, the chapter addresses the SRMP practice that I codeveloped with William Brent, giving particular emphasis to the way that it nominates sound as a paradoxical relationality that reaches toward the "fieldness" (in McLuhan's sense) of this relationality precisely by refusing to give sonic instantiations (i.e., sounds) primary status.[39] In this context, embodied organisms do not overdetermine their representations (as Hansen would have it) but rather coexist with them via complex intermediating networks. Ultimately, then, SRMP models a way in which sound disjunctively intervenes in constructions of presence and absence, opening its practitioners to a relational play that not only moves between those two poles but also constructs them as poles (even as it is constructed by them).

Inevitably, this text falls prey in advance to the danger that it seeks to confront; namely, in the effort to reverse the flow of theory to practice, my use of (relatively) rational argumentation and conventional language as the medium of presentation renders my effort a paradoxical one. In response, I can only insist that my aim is not to construct a metaperspective from which theories of posthumanism can be considered but instead to investigate the flows and intensities that come to the surface when an effort is made to hold competing perspectives in tension with one another. To this end, the concluding chapter of this book focuses less on summarizing the text's claims than it does on emphasizing their contingency: if the body of the text—especially the theoretical chapters—tends to account for its fields of study in descriptive language, the conclusion reiterates the extent to which the rhetoric of this approach masks its own inevitable biases. In this sense, the conclusion reminds the reader that though the text describes three technologies of posthumanism, it also performs a fourth.

PART I

1 FROM GENES TO MEMES: OLLIVIER DYENS AND THE SCIENTIFIC POSTHUMANISM OF DARWINIAN EVOLUTION

> Thinking, no doubt, plays an enormous role in every scientific enterprise, but it is the role of a means to an end; the end is a decision about what is worthwhile knowing, and this decision cannot be scientific.
>
> HANNAH ARENDT, *THE LIFE OF THE MIND*

> Were he immortal, an existent would no longer be what we call a man.
>
> SIMONE DE BEAUVOIR, *THE SECOND SEX*

In this chapter, I elaborate a notion of technological posthumanism that is predicated on understanding information—specifically replicable data—as the dominant term in relations between culture and technology. To exemplify this perspective, I focus my argument on the theoretical work of Ollivier Dyens, particularly his book *Metal and Flesh*. To begin, the chapter emphasizes Dyens's redefinition of the body as a cultural entity, drawing out both the connections and points of departure that this understanding shares with Richard Dawkins's (in)famous "selfish gene" argument and particularly focusing on this perspective's rendering of life in terms of probabilistic functions. Having thus established the terrain of engagement, the chapter proceeds to consider this construction technologically, going in search of its hauntologies to ask what is foreclosed by the positive definition of life espoused by Dyens and Dawkins. From the doubled movements of this inquiry (where life is both a subject and an object of analysis), the chapter argues that Dyens's posthumanism—precisely in its constative claims to scientificity—performs a deconstruction of scientific materiality, namely, the claim that scientific data are measurable, repeatable, and falsifiable. Specifically, *Metal and Flesh*'s deconstruction of the human slides back to deconstruct the scientific tenets of evolution itself:

scientific posthumanism is thus a site where science is no longer scientific and where the scientific presumptions that underwrite Darwinian evolution are desublimated.[1] However, since Dyens's thought is itself predicated on the truth claims of this scientific frame, I ultimately argue that scientific posthumanism is trapped within a nihilistic framework of its own making.

Why Ollivier Dyens? There is, of course, the content of his work, which has made important contributions to the field of technological posthumanism: his presentations of a robust selection of posthuman scholarship in the forms of a Google Earth skin[2] and a multimedia web database,[3] for example, are all the more important considering the tendency of research in this area to proliferate through diverse channels (even more so than in conventional scholarly disciplines). Moreover, Dyens's contributions as a digital poet nominate him as someone whose theoretical writing connects to a mind that is sensitive to the vagaries of artistic practices. Most important to this project, though, is Dyens's intensification of Dawkins's reading of Darwinian evolution, a trait that distinguishes Metal and Flesh as a promising place to begin deconstructing those principles' implicit values: what unmarked privileges are hidden in the effort to uncover laws of reproduction? Finally, while Dyens's work is peripheral to the discourse of evolution—in part because Metal and Flesh flouts many of the conventions of scholarly writing—this marginality is an asset inasmuch as it denaturalizes the apparent neutrality of evolutionary principles, revealing the scientific claims to neutrality that ground evolution's narrative.

With respect to this latter point, an important thread throughout this chapter is the observation that the materially nonprogressivist stance of science belies a profound discursive progressivism. Specifically, science's refusal to identify change with progress in the phenomena it observes is undercut by its larger-scale faith in a progressive expansion of human knowledge. In this context, Dyens's deployment of evolution exposes the tension between science's recognition that progress is only one way of accounting for phenomenal changes and its larger insistence that all discoveries contribute to (a narrative of) human progress. Once visible, this tension itself undercuts the claims to neutrality of Darwinian evolution so that Dyens's thought turns back to rewrite itself as having sublimated its own contingency.

THE PRIMARY OBJECTIVE of *Metal and Flesh* is to redefine the human body as a cultural entity, and this redefinition hinges on two key observations: first, that survival and reproduction are processes that take place at a level below that of the species or individual, and second, that these processes—in the context of ubiquitous media technologies—have detached from the genetic biology that had formerly determined them. In short, Dyens argues that humans are the by-product of reproductive processes that have historically been represented genetically but that are now represented culturally in the form of information (broadly understood). Simply put, if all bodies survive by replicating, *cultural bodies* are bodies whose reproduction is dominated by nongenetic replication.

Fundamentally, this notion of cultural bodies is an extrapolation of Dawkins's populist argument that genes operate "selfishly." In its most basic form, Dawkins's theory reiterates Darwin's insistence that evolution is an essentially genetic process: genetic traits that—for whatever reasons—optimize longevity and reproduction are more likely to remain active in the biological economy. For Dawkins, this is emphatically not a moral or ideological claim but simply a description of fact: what is present today indicates the things in the past that were most likely to survive. As a result, Dawkins repeatedly insists that genes do not behave "selfishly" in a subjective sense but rather in a behavioral sense; that is, they have the *effect* of improving their own survival prospects. The common mistake, Dawkins argues, is to think that our subjective intentions are somehow detached from our genetic biology, which is to say that the mistake lies in thinking that our will—expressed through categories of thought—is the cause of our embodied actions. In contrast, Dawkins advocates acknowledging thoughts themselves as a means of genetic survival.

This aspect of Dawkins's thought has proved to be fecund material for numerous and diverse scholars over the last forty years and has contributed—along with the related inter-, extra-, and multidisciplinary work on "complexity" carried out at the Santa Fe Institute (since 1984) and elsewhere—to a migration of disciplines like cognitive science to questions that have traditionally been the province of the humanities. For example, a recent post by John Doris to the National Humanities Center's "On the Human" forum cites a variety of studies that each suggest, in different ways, that a sizable portion of our justificatory thinking (with respect to decision making) is done only after a decision has already been made. In the most striking such example, Doris recounts a study in which participants consistently failed to detect mismatches between intentions and outcomes in a simple decision task: they were asked to make a decision of preference between two pictures of human beings but were subsequently offered (through sleight of hand) the image that they did not choose *as though it were the one that they did*. Interestingly, the switch was detected less than 26 percent of the time, a rate that remained consistent regardless of the degree of similarity between the paired images (though, to be fair, the varying "degrees of similarity" were constrained to "normal," "healthy" humans). Moreover, when asked to explain their choices, the explanations of the duped participants (who were still unaware of the sleight of hand) were effectively indistinguishable from the explanations of the control group: there was no indication "such as evidence of deceit or hesitation, to differentiate the reasons participants [gave] for the choices they did make from the reasons they [gave] for the choices they didn't make."[4] This leads Doris to conclude that "though they didn't know what they did, they had no trouble coming up with reasons why they did it."[5]

For humanities scholars, this example may suggest an operation of the Freudian unconscious in which the experiment's conductor is lent an authority akin to that of a "subject presumed to know."[6] In Dawkins, though, such findings were anticipated in a different register through his emphasis on the formal operation of replication, which he takes to be "the ultimate rationale for our existence."[7] In this view, we must first understand individuals as the "survival machines" of replicators; genes are the most relevant type of replicator because they are the smallest unit in a nested hierarchy and because they are the most abundant decipherable division of biological matter. That is, because genes are subsets of cells, which are subsets of individuals, and so on, they represent the stratum of biology capable of the most influential and nuanced interactions. Thus the point, for Dawkins, is that evolved characteristics in species and individuals come about as the result of successful genetic *reproduction* rather than individual species' *mutation*.[8] Genes are said to express themselves, then, and their successes in doing so are registered as characteristics: it isn't that opposable thumbs, for example, evolved to allow us to text message more efficiently but rather that human bodies were the optimal carrier for the opposable thumbs genetic combination—itself the most optimal carrier for its constitutive genes—as well as the optimal carrier for whatever genetic combination is expressed in text messaging. For Dawkins, genes are thus constitutively "selfish" simply because they are the basis for the complex hierarchy of biological life, including the much-vaunted decision-making skills that inform our actions.

However, Dawkins begins to trouble this neat and tidy picture of evolution in the final chapter of *The Selfish Gene*, where he notes that biology is not the only medium of evolutionary processes. Citing in particular the "analogy between cultural and genetic evolution,"[9] he further remarks that it even appears that culture is "achieving evolutionary change at a rate that leaves the old gene panting far behind."[10] As a result, Dawkins insists that, while "DNA has been the only replicator worth talking about [for more than three thousand million years], it does not necessarily hold these monopoly rights for all time."[11] It follows, then, that a robust account of evolution cannot foreclose the possible emergence of new types of replicators.

Dawkins coins the term *meme* to account for such replicators as they are found in the cultural realm, a nomination that specifically designates a "unit of cultural transmission, or a unit of imitation."[12] A meme might manifest in any number of forms, ranging from a fragment of a symphonic theme to an idea, a fashion, a way of cooking rice, or virtually anything else; the key point is that memes propagate in the meme pool by "leaping from brain to brain *via* a process which, in the broad sense, can be called imitation."[13] As such, memes in a meme pool are equivalent to genes in a gene pool so that the two forms of replicators compete with one another for evolutionary superordination. In the context of Dawkins's larger project of evolution advocacy, the point to note here is that—somewhat paradoxically—the existence of these nongenetic replicators actually strengthens his overall genetic argument because they naturalize the evolutionary process (specifically, a hierarchically structured reading of Darwin's theory of natural selection) as a process rather than as an ontological claim.

Thus memes mark Dyens's point of entrance into Dawkins's thought because they allow for constitutions of life that are entangled with biology but that do not necessarily spring from it. That is, Dyens's cultural bodies merge ideology and biology such that *Metal and Flesh* postulates "an ontological essence for the body as a semiotic interface to the complexity . . .

of information ecosystems."[14] As a result, though genetics and memetics each unfold in distinct ontological strata, they nonetheless combine via their shared reliance on informatic reproductive processes (informatic in the sense that they are taken to be composed of discrete units of data, be they biological or cultural). From the perspective of cultural bodies, then, technology, culture, and biology are simultaneously distinct and inseparable from one another.[15]

Important to this perspective is a definition of culture that moves away from the term's conventional usage toward the sense captured by Dawkins's notion of memes. Accordingly, culture, for Dyens, is "any piece of information that can reproduce and disseminate without making direct use of genetic channels."[16] Moreover, culture is not limited to human activity but instead includes "any trace left in the environment by a living being."[17] Employing these two principles in tandem, then, Dyens's notion of culture includes books, songs, and myths (for example) but also nests, scents, and the flight paths of migratory birds. To say that bodies are cultural is thus to describe them as entanglements of multiple, autonomous strata that each behaves according to its own particular logic, a perspective that aligns with Hayles's description of her subjective agency as the site at which multiple agents collide.[18] Although this assertion has obtained to some extent throughout history, the current ubiquity of media technologies makes it especially true for our present era. Moreover, this quantitative shift is accompanied by an equally compelling qualitative shift in the realities that media convey, leading Dyens to note that because today's technologies "offer us access to new levels of reality that our biology cannot perceive, define, or understand by itself,"[19] the intensity of cultural bodies' frequent encounters with simultaneous levels of reality that are constituted through contradicting absolutes is significantly higher than it has historically been (and is continually increasing).

Dyens's emphasis on the intensification that takes place in contemporary technoculture leads him to argue that the accent in *cultural bodies* has now shifted from embodiment to culture. That is, Dyens argues that memes proliferate cultures but that these cultures are qualitatively changed by this multiplication; human cultural artifacts, for example, are no longer definitively human and are thus no longer even artifacts (in the proper sense) but instead take on a life of their own. Moreover, this logic feeds back to reconstruct humanity: the conventional assignation of mastery to life over objects is dramatically reconfigured when objects themselves become alive. In this respect, Dyens echoes visions of technological society offered by Arthur Kroker and Langdon Winner, where the former argues that individuals are possessed by virtual reality[20] and the latter promotes "technological somnambulism" as the way to understand our ignorance of the fact that we are being remade from the inside out.[21] In effect, Dyens argues that "culture saps the biological environment not because it is intrinsically negative or ill-willed, but because it gives replicators the ability to bypass organic matter and biological channels."[22] Indeed, Dyens even goes as far as to suggest that the current state of environmental crisis is a result of this transformation of the biosphere.[23]

Consequent to this perspective, Dyens follows Dawkins in registering life in the positive terms of a probability function rather than as a categorical distinction between life and death. That is, life is defined (in *Metal and Flesh*) in terms of what persists, and what persists is a function of how life reproduces itself. Rather than existing within the category of either life or nonlife, then, the life of cultural bodies lives on a continuum. That is, not only does "the structure that allows us to positively differentiate between life and non-life"[24] elude us but it eludes us precisely because nonlife is excluded from the definition of life, which registers living-ness on a continuum based on the probability and rate of genetic or memetic replication.

Because such a continuum is defined by interactions (i.e., rates of replication and survival) rather than categorical distinctions, it follows that it does not really pertain to life and death per se but is rather an indication of a particular organism's degree of complexity. As Dyens explains,

> the implication of this morphing of the human body into a cultural one is that no living body can be singled out. Each body is several living ones simultaneously . . . so that phenomena seem to distinguish themselves according to their degree of complexity, not according to absolutes like life and death.[25]

Crucially, then, life is divested of any de facto value or identity and is instead symptomatic of complex interactions (between genes, memes, or both).

Clearly this construction of life risks tautology when it is considered alongside the evolutionary theory from which it sprang: life is defined in terms of the ability to survive (through replication), but survival itself is divorced from any necessary ties to organic life and is measured instead—post factum—as that which has persisted through the process of evolution. That is, life is that which persists over time in an economy of replication, but replication itself is considered a fundamental activity of living so that we are left with the intention–outcome experiment that Doris cites, without even the alibi of having chosen (because life is naturalized).[26] This line of critique has, of course, been leveled at Darwinian evolution more generally: for example, during a recent lecture at the Université du Québec à Montréal, Jerry Fodor indicated tautology as a crucial flaw in the Darwinian perspective because it means that evolution does not support counterfactuals (i.e. "this is what would have happened if that had happened") in the way that, for example, Newton's $F = ma$ does.[27] Furthermore, as Stevan Harnad points out in his response to Fodor, this criticism is not usually even disputed by Darwinians, who are apt to argue that Darwin's contribution is to offer a perspective that—in being categorically true—yields crucial *methodological* consequences[28] rather than optimizing any particular (mechanical) effects.[29] The crucial point, in the context of the argument being presented here, is that both supporters and detractors of Darwinian evolution agree that tautology is (at the very least) a proximate concern. Without entering the debate regarding the facts of Darwinian evolution, then, we can observe that perspectives that result from this understanding—perspectives that seek to turn these methodological consequences into optimized mechanical effects—have at their core, necessarily, a claim to originary and unambiguous truth. That is, Darwinian evolution may not be methodologically tautological in its descriptions, but extrapolating its principles to a predictive or categorical framework—that is, to define what life is now or soon will be—is an activity that is either tautological (in that its very essence reveals itself to be true), predicated on a naive objectivity (in that its truth claims are expressed through the "neutrality" of an observer), or a mixture of both.

From this observation, then, we can construe evolution in its categorical and predictive framework as always-already separated from its etymology as a verb—the Latin *evolutio*, for "unrolling," from *evolvere*, "to *become* a simple statement of fact," namely, that that which exists, exists. Dawkins is sensitive to this danger and attempts to evade it by insisting on the role of empirical data in selfish gene theory. Thus, for example, he offers the following in *The Selfish Gene*:

> If you look at the way natural selection works, it seems to follow that anything that has evolved by natural selection should be selfish. Therefore we must expect that when we go and look at the behavior of baboons, humans, and all other living creatures, we shall find it to be selfish. *If we find that our expectation is wrong,* if we observe that human behavior is truly altruistic, then we shall be faced with something puzzling, something that needs explaining.[30]

In the italicized section, Dawkins fuses the objectivity of empiricism with the subjectivity of expectations and predictions, rhetorically promoting the claims to neutrality of the former onto the latter. Indeed, Dyens's intensification of memes demonstrates the speciousness of this argument because its posthumanist orientation results from memes not being subject to empirical data collection in the same way that genes are (because their material boundaries are less clearly delineated).

The question, then, is not whether there is a tautology but rather where this tautology is located and what it services. Put differently, what is the technology of a tautological definition of life? The immediate answer to this question is that life becomes, for Dawkins, a second-order simulation, where the distinction between "life" and "life itself" begins to break down.[31] That is, Dawkins constructs subjective agency as an exclusive function of genetic processes (i.e., agency is simply a genetic survival mechanism), which is to say that subjectivity is completely determined by genetic activity. Indeed, any *mobilization* of selfish gene theory—that is, any movement of the theory beyond a simple statement of fact—must account not only for subjective agency (as Dawkins does, by positioning it as an offshoot of genetic activity) but also for the particular overdeterminations that spring from each specific narrative iteration of this theory. That is, though Dawkins's recourse to the objective materiality of science sidesteps (to an extent) the risk of tautology, this sidestepping itself instantiates a version of science that is lent a narrative authority that directly undermines this objectivity, since it is now subjectively oriented (i.e., it is a narrative). Indeed, Dawkins seems to be attuned to this slipperiness, which leads him to repeatedly insist that the grammar of terms such as *survival mechanism* is actually misleadingly subjective because *mechanism* implies a genetic agency rather than a probabilistic function. However, the fact remains that this grammar is necessary for his observations to narrate, which in turn is necessary for them to speak outside of their tautological dimension. Thus, if Dawkins is untroubled by questions of subjective fatalism (which his emphasis on the amorality of selfish gene theory strongly suggests), this may simply be because he fails to avow the technological component of his own thought.

In this context, Dyens's movement from genes to memes can be read as a corrective of this failing: memes set the trajectory for Dyens's departure from Dawkins as much as they mark his entrance into Dawkins's worldview. In this sense, *Metal and Flesh* charts the migration of life from second- to third-order simulation, where the process of life precedes its instantiation and precludes recourse to a biological real. That is, by extending the language of biological probability (through which Dawkins renders individuals) to include "cultural" human behaviors, the highly specified language of genetics paradoxically multiplies. Because these behaviors have their own languages and metrics of survival, survival itself is reconfigured in a way that undermines the specific examples through which it was initially defined; the (abstract) reproductive process of survival is privileged over any particular instance of it and over any identity produced by it. As a result, the question arises: what happens when memes no longer operate analogously to genes but instead act in tandem with genes in relation to a third term, the privileged process of replication? Moreover, in the context of a "mediascape"[32] that convincingly naturalizes memes, is it possible that memes have come to dominate genes even within their pairing? If this is the case, what new forms emerge from these newly contingent genes?

TO DESCRIBE DYENS'S CULTURAL BODIES in the preceding terms is already to begin to think them technologically in that the memetic process that produces cultural bodies is an "extension" of the genetic technology that grounds Dawkins's narration of biological bodies. What, though, of the remaining three intensities of McLuhan's fourfold? To understand cultural bodies as technologies of (posthumanist) subjectivity, we must also ask after their reversals, obsolescences, and retrievals. This is the case because, though cultural bodies are bodies produced by technological culture, they are also themselves technologies.

The basic principle of Dyens's thought is, I think, irrefutable. Western cultures derive from a particular physical conception of materiality that is ultimately untenable in light of the scales of reality that have been ushered in with new technologies. In the broadest and most basic sense, our foundational laws are simply unable to account for present realities, a fact that is seen equally in the interminability of debates over file sharing and stem cell research (i.e., the fact that the debates themselves seem to have no end forthcoming) as it is in the persistent deconstruction of everything from gender to disciplinarity. Quite simply, subjectivity—along with the collective and individual norms, laws, rituals, and so on, that connect to it—is deeply entwined with the specificities of media (which are themselves alive, according to Dyens's formulation), an idea that Mark Poster has dramatized in his discussion of analog subjectivity and print culture.[33] Moreover, in the context of numerous arguments similar to Poster's, it is reasonable to conclude that various forms of "law" are imbricated in one another so that laws of physics (for example) are connected to juridical laws.[34] Indeed, the current rash of global crises—economic, environmental, and political—might all be attributed in part to an incommensurability between the (normative) laws that govern us and our reflection of them through our daily practices, and this incommensurability directly relates to the profusion of scales that are now in play.

What Dyens raises, though, is an understanding that places normative laws on the same ontological stratum as the actions that they direct. As a result, actions are not actually governed by normative laws in his view; instead, the laws themselves are understood as particular actions that interact, nonhierarchically, with other activities. In this respect, scientific posthumanism relates multiple realities to one another through points of intersection and fields of overlap rather than (for example) through a shared relation to an (extradiscursive) Real. Correspondingly, conflicts between various strata of reality result not (for Dyens) from an originary lack but from an excess of data that overwhelms attempts to construct linear causality wherever such scales coexist.

Cultural bodies, then, constitute an attempt to realign these disparate strata by articulating them in the dematerialized (or "a-medial") terms of information. We have already seen how this plays out with respect to the constitution of life, where life is redefined as a continuum of intensity rather than as a category of inclusion–exclusion. Similarly, Dyens describes a perpetual reconstitution of realities that takes place through the shifts in perspective afforded by contemporary technologies.[35] As with life, though, these realities do not correlate via the category of reality (or the Real) but rather through the very process of reality construction that yields them. In this respect, life and reality feed back into one another because the perspectives offered by new technologies are defined through the content that they yield in the conventional domains of living beings (i.e., the scales and intensities of reality that they make visible). As a result, explosions of life and reality are always-already implicated in one another, with each participating in the process of the other (i.e., realities themselves participate in a process of living, and vice versa) because they are both understood by Dyens as dematerialized processes (in the sense that they are culturally overdetermined).

This entanglement of reality and life is most evident in Dyens's discussions of virtual reality. For example, Dyens contradicts media artist Char Davies's claim that virtual beings have "no otherness" to argue, in contrast, that the mystery of virtual beings is precisely the coevolution of man and machine of which they are the conduit.[36] However, though Dyens takes this to mean that a virtual being is "a perception come alive,"[37] it is more the case that this is a conflation of life and a particular stratum of reality (i.e., virtual reality). That is, *evolution* in this formulation does not signify a development in humans or machines but rather the emergence of a perspective that they both might share. As a result, this account bolsters the conception that the boundaries of life and reality are simultaneously breached and reinforced through recourse to the other.

This simultaneity, then, reveals a paradoxical causality at the core of Dyens's construction of life, a code that operates as its constitutive ambivalence. On one hand, life is an emergent phenomenon of cultural (including genetic) interactions, suggesting that it is an (unpredictable) effect of (not necessarily living) processes. On the other hand, though, life is also "an energy . . . that simply uses the forms and materials that are most useful to it,"[38] implying that living beings possess a form of agency that distinguishes them from their materials.[39] This paradox is not, in and of itself, particularly notable, but what *is* notable is that it is not avowed as a paradox by Dyens: in the same moment that the term *life* explodes a notion of living to multiply the Real into an infinite play of realities, it also congeals the ambivalent causality through which this multiplication takes place. As a result, the migration of evolutionary processes between biological and cultural strata is concealed, and the evolutionary process itself is naturalized. In short, by registering constructions of life and reality on a shared continuum of evolution, the leap from evolution's biological foundation to Dyens's cultural logic is sublimated.

From this perspective, we can reinvigorate Dyens's guiding question of what happens when memes become the dominant term in the memes–genes relation: in the coproliferation of life and realities—a movement that extends life beyond the language of exclusion to include all possible entities—what is changed? As an immediate response, this recalls the informatic reduction of embodiment that is regularly the subject of Hayles's critique and that is considered in chapter 3 of this volume: we might say that Dyens's notion of life obsolesces questions of subjective ethics and individual embodiment because human-scale reality is no longer any more important than any other stratum of reality.[40] Moreover, if we accept Dyens's argument that we are now dominated by culture, we are also forced to accept that the values of biological embodiment are no longer primarily determinant.

This forced disappearance of embodied values suggests a host of further obsolescences, situating Dyens as a successor to the recent history of French thinkers about whom Kroker remarks, "In their collective imagination is rehearsed the terminal symptoms of the age of technology triumphant: the death of politics, the death of aesthetics, the death of the self, the death of the social, the death of sex."[41] Quite simply, Dyens's narration of technology's cultural materiality performs a (by now familiar) poststructuralist disappearance of universal values. And yet, what nominates Dyens as an author of *scientific* posthumanism is the fact that he consciously distances himself from this poststructuralist perspective through recourse to the scientific provenance of Darwinian evolution and further argues that one must ultimately choose between what he

characterizes as the vagaries of French thought (citing specifically Deleuze) and the more immediately material realities of science.[42] In a
sense, then, the body is doubly obsolesced in Dyens's thought: first, the
particularities of flesh disappear in the positive language of science that
renders them as probability data, exchangeable with cultural replicators;
and second, this disappearance itself is disavowed through his insistence
on the objective truth of this rendering, which is presented as though it
were free from bias and subjective influence. As a result, it is irrelevant
that Dyens does not see himself as an advocate of posthumanism in its
transhumanist appearance, where technological progress is taken to be an
ethical imperative.[43] In fact, Dyens's advocacy runs much deeper because
it is spoken in a language of inevitability.

THERE IS A TWIN MOVEMENT to Dyens's thought: on one hand, it moves
away from Dawkins's genetic determinism by extending genes beyond the
metrics of biology; on the other hand, this movement away reinscribes
precisely the logic that drives genetic determination. In this sense, the
language of science—the injunction to measure, repeat, and theoretically
falsify—constitutes the reversal potential of both cultural bodies and Darwinian evolution: the unmarked values of science are reinforced precisely
in the cultural body's movement away from conventional conceptions of
material reality.

   With this in mind, consider Dyens's argument that there is a *natural*
biological attraction that has been subsequently overtaken by culture.
This shift, he argues, occurs because the metric of fertility is changed by
the ubiquity of media. In short, "human beauty is pulled from its foun-
dations"[44] because our contemporary media environment results in our
seeking culturally (rather than biologically) fertile bodies. As a result, a
sex-symbol celebrity such as Pamela Anderson was in the 1990s (for example) is no longer precisely human but is instead a "network of signs
and desires [whose] ontology is cultural in nature."[45]

   However, this raises the question of whether human beauty was ever
actually primarily biological, as Dyens suggests. That is, has desire ever
been simply about reproduction? Theorists as diverse as Lacan, Foucault,
Deleuze, and Butler (to name only a few) have argued persuasively that
this is not the case. In response, Dyens offers the historical consistency

of the "ideal waist-to-hip ratio" in "beautiful" women as evidence, but this contention neglects that the term that grounds the argument (i.e., *beauty*) is itself discursively charged, specifically as that which paradoxically sustains the nature–culture divide that it bridges.[46] Beauty thus becomes a slippery metaphor in Dyens's hands, simultaneously reified through its relation to biology while remaining necessarily abstract in its cultural deployment (because to mean at all, beauty has to mean something beyond simply reproduction).[47]

In Dawkins's case, he responds to this line of critique by redoubling his emphasis on complex interactions that take place at the genetic level. That is, implicit in his argument is the idea that cultural values themselves exist only relative to genetic replication, which is to say that values that seemingly interfere with the ability of individuals to reproduce might nonetheless persist because they are useful to the replicative tempi and processes of certain genes. The key, in this perspective, is that genes occupy a privileged position in the gene–behavior relation such that behaviors are symptoms (rather than causes) of genetic processes. In this sense, Dawkins's theory is profoundly anticonstructivist, but this is not to say that he denies any link between behaviors and genetic processes. Instead, by treating perceptible processes symptomatically, Dawkins simultaneously severs and reinforces the bidirectional interaction between the two strata. That is, behaviors (as symptoms) are relegated to being a means of registering and measuring genetic processes; however, precisely as such, they dictate the ultimate terms through which the latter are understood, specifically delineating the meaning of genetic success. By denying the material consequences of this doubled move, Dawkins is bound to a scientifically objective notion of materiality that ultimately confines his thought to a human scale of perception (though, to be clear, he does not acknowledge this). Notably, this is not because he is limited by human sense organs (which his emphasis on genetics moves beyond) but because his narratives of genetic processes are continually constrained by the necessity to make sense as human interpretations.[48] Moreover, they are interpretations that ignore the material differences between strata of reality so that Dawkins's claim to genetic materiality has the character of alibi: while appearing to be enactors of evolutionary activity, genes in fact function in Dawkins's narrative to distract attention from the narrative element of evolution itself.

This is the sense in which Dawkins's seemingly anticonstructivist position does not actually engage with its purported interlocutor. That is, the constructivist position does not consist in denying the existence of material facticity but rather in the assertion that entrance into discourse always-already marks specific facts. As Butler has famously argued with respect to sex, to concede the undeniability of a given fact is always to concede a particular valorization of it so that the discourse through which that concession inevitably occurs is "formative of the very phenomenon that it concedes."[49] To claim that a particular utterance is true, then, is also to perform the reality (or, in Butler's case, the subject position) whereby that is the case. If a different subject position were performed (which is not always, if ever, a matter of choice), the criteria for truth would also be shifted.

In this context, it is clear that Dawkins's argument masks the assumptions it is predicated on so that, ironically, we might accuse him of anthropomorphizing: in his insistence on the genetic basis of human behavior, Dawkins veils a much deeper attachment to certain cultural sexual mores, namely an economy in which survival and reproduction are part and parcel of the same intensity.[50] This attachment naturalizes both the equation of survival and reproduction and the privilege that he affords them, which in turn insulates his reading of Darwinian evolution from the accusations of tautology discussed earlier. As we saw with reproduction and beauty, *survival has to mean something beyond simply persistence*, and genetics is not the territory where this meaning is registered (i.e., where it would mean). Indeed, this is a frequent criticism of science that has arisen within that discipline's reflexive arm: as Hayles has noted, "one of the important insights that has emerged in science studies in the last 20 years is the realization that scientific models are underdetermined with respect to empirical evidence (or, to put it another way, that multiple models may be consistent with prevailing knowledge)."[51] In this light, Dawkins's claims to objective truth based on such empirical evidence are not only paradoxical (as discussed earlier) but also insufficient.

Metal and Flesh is haunted by a similar problem, though it is arrived at from the opposite perspective. Consider, for example, Dyens's argument that "human beings . . . are nothing more than a specific historical construct."[52] What value is indicated by "nothing more than" in this statement?

Is this to suggest that there exists something that is not historically situated? At the very least, the statement suggests that understanding humans as epiphenomena detracts from the value that they would possess if they were not. In this comportment, even if Dyens does not posit the existence of a *neutral* transcendental universal, he nonetheless reserves a space for a hegemonized one. As a result, we might say that Dyens acts in the manner of a bourgeois individual, in Žižek's sense: he *thinks* that the Universal is a property of the Particular but *acts* in the opposite way.[53] In this way, what Dyens misrecognizes is not reality (i.e., the fact that humans are epiphenomenal) but the illusion that structures that reality (i.e., the value he attaches to a notion of "humanity" that is *potentially* transcendental and universal). The illusion is therefore double: it consists in overlooking the illusion that structures his effective relationship to reality, in this case, the implication of value itself in the very language that speaks against particular values. In short, we might say that subjective meaning—as a performance of value—functions as Dyens's ideological fantasy.[54]

Stated succinctly, then, this is to say that it is necessary for both Dawkins and Dyens to posit *reproduction* and *survival* as terms that are privileged a priori. Importantly, it is not the particular content that is determinant here but the positing itself. In this light, it is not the coexistence of heritable traits with random alterations that is the essence of their reading of Darwinian evolution; instead, these discoveries simply reinforce the injunction to *mean* (the necessity of narration) that has been a mainstay of humanity throughout its (necessarily narrated) history. In this respect, the evolutionary assumptions of scientific posthumanism are complicit with the very humanism against which scientific posthumanism was mobilized in the first place.[55]

Consider, then, that Dyens purports to think the ground zero of the human–technology nexus outside of conventional anthropocentric language. Following McLuhan, *Metal and Flesh* is predicated on the notion that we perceive reality through a rearview mirror so that our natural and intuitive worldviews are geared toward what has already happened rather than what is taking place in a present moment. As a result, we can only perceive our present realities, and make choices about our future directions, by examining the technologies through which we perceive the world rather than by attempting to directly encounter the world itself. In

Dyens's case, he pursues this task through his insistence on the paradoxical and entangled status of perceptions derived from the multiplicity of technologies that are now active.

However, as has been alluded to, the fundamental influence that technology has on our perceptions today may not reside in the perceptive apparatuses of these technologies themselves but instead in the disavowed value system that is activated through them. As McLuhan always insisted, media do not simply project their purported "content" but also project their medial biases onto this purported content. Thus, to probe the technology of scientific posthumanism, we might ask after what is achieved through the privileging discussed earlier; that is, what specifically is achieved by Dyens's culture-based construction of "survival"? In its simplest form, the answer to this question is another sense in which Dyens's posthuman is scientific: in lieu of recognizing human individuals as the source of a relatively limited range of perceptions, Dyens transcends this limit by emphasizing the multiple processes at work in any given perception. Through this emphasis, otherwise incommensurable strata can be compared: memes and genes, for example, but also humans and machines. Dyens's posthuman is scientific, then, because its economy consists in that which can be registered scientifically: pure scientific reasoning, detached from any specific biology, is the technology of those levels of reality that elude our conventional sensory apparatus but are nonetheless quantifiable. That this quantifiability manifests in *Metal and Flesh* as that which eludes empiricism does not rob Dyens's argument of its scientificity but rather pulls it closer to the heart of science itself.[56]

To understand why this is scientific, consider the obvious question that is never asked in *Metal and Flesh*: what of the imagination? That is, if contemporary media have rendered culture material and ideas alive, what is the status of the imagination? Do hallucinations occupy the same ontological territory as memes? Is nothing now falsifiable? These questions do not enter into either Dyens's or Dawkins's accounts precisely because they cannot possibly register there: in both cases, the imagination can only enter into consideration when its otherness has been sublimated to a process of replication. Emphatically, then, the movement of evolutionary processes into the cultural domain must be—in Dyens's account—unilateral. As such, the extension of the definition of life to include viruses, ideas, and machines is a misnomer. Instead, it is the nonliving and disembodied language of science that has been extended to what was the realm of the living but is now rendered exclusively as a "badly analyzed composite" of probability data.[57]

Simply put, cultural bodies are dematerialized bodies. In them, everything conceivable becomes relatable, and everything relatable becomes interchangeable. Energy is disentropized because it is made a matter of exchange, a matter of income and expense in a static economy of replication: nothing living, after all, escapes Dyens's unary code of survival. In this sense, cultural bodies are a completion of biological science, taking form in and as the moment that biology finally dispenses with the alibi of fleshly corroboration in favor of the quantifiable materiality of the discipline of science. In cultural bodies, evolution is no longer a scientific theory—is no longer a theory that is unified, repeatable, and falsifiable—but instead occupies the position of scientific materiality itself, namely, that through which what exists is registered as such.

Again, though, it must be emphasized that this dehumanizing dematerialization is also a hyperhumanization (although Dyens does not present it as such). The genealogy of science, after all, is profoundly anthropomorphic in that it situates knowledge within a progressivist narrative of accumulation (in the sense that scientific discovery is predicated on the assumption that such discovery moves one closer to an objective truth of the matter at hand).[58] There is a sense in which cultural bodies are, after all, human bodies because in them is performed the human's redefinition of itself, where the obsolesced human individual (obsolesced because the individual is no longer a sensible unit of distinction for Dyens) returns to haunt scientific thought as its grounding metaphor. Thus a correction to the observation that the movement of scientific measurement into life is unilateral (as a probabilistic construction of life suggests): this unilaterality bespeaks a paradoxical twin movement—always evading scientific consciousness—of life into science.

The scientific posthuman is a simulated human, then, but it may be all the more human for this fact. That is, it exists in a logic of simulation, where the thing initially taken to be indicative of an outcome becomes the outcome itself. However, because Dyens aims to escape the language of simulation, his failure to accomplish this is—paradoxically—also his success. That is, Dyens's failure might be read as a paroxysm of the evolutionary code that rules both his and Dawkins's thought: his mistake, to put it crudely, is to fail to avow this failure as such and to take refuge instead in a language of causal reason.

Thus Dyens's cultural bodies cannot ultimately account for the very shifts in perspective of which they are taken to be indicative. In fact, this is seen in the claim in *Metal and Flesh* that cultural bodies are ultimately bodies made of representations wherein "linearity and causality no longer exist"[59] so that "each body is several living ones simultaneously."[60] As mentioned earlier, because demarcations are a matter of perspective, phenomena are distinguished relatively, according to their degree of complexity (rather than absolutely, according to categories of life and death). However, the question remains: where is this complexity viewed from so that it can register as such?[61] That is, if complexity exists in measurable degrees and is oriented in a causal topology, doesn't this metric itself amount to an (hierarchical) abstract category that quantifies qualitatively different phenomena? In this sense, complexity—as a metric—is a transcendental category, and one that always performs more than it claims: again, "greater complexity" means something because it appeals to something beyond its avowed purview so that the deterritorializing movement of complexity is reterritorialized in this appeal.[62] As in Dawkins, this suggests that *Metal and Flesh* glides between expressing life as multiplicity (i.e., life as composed of qualitative differences) and life as multiple (i.e., life as composed of numerous iterations of a single essence) without fully attending to either understanding.

Dyens seems to be intuitively aware of this discrepancy in his suggestion that representations are necessary to existence because they act as time buffers.[63] That is, there remains the sense of a locatable, unified force that is active in this theory (in this instance, housed under the concept of time), despite the rhetoric of nonlinearity and a-causal relations. While it is true that this impetus is obnubilated by changes in name—from "time" to "life" to "evolution"—its tautological configuration remains constant throughout. As a result, we might argue that the ultimate reversal potential of the process of dematerialization that yields cultural bodies is the rematerialization of experience that was the project of Deleuze and Guattari. That is, by remaining trapped in the simulating logic of

semiotics, the cultural body performs the possibility of "a rhizomatic
network of experience where events vanish into a decoded world of immateriality, only to instantly reappear in their opposite sign-form in an
endless chain of 'lines' of flight and interruption."[64] In Dyens's failure to
address this possibility, it is paradoxically made more proximate as the
completion of cultural bodies' semiotic performance.

WITH ALL OF THE PRECEDING IN MIND, it seems clear that the territory
from which cultural bodies spring is not only the nexus of humans and
machines (as claimed) but also (and more fundamentally) a scene of language. In demonstration of this, throughout this chapter, I have intentionally conflated the theories of Dyens and Dawkins, with the latter also
liberally mixed with Darwin.[65] Through this tactic, I have (performatively)
indicated the slipperiness of the metaphors employed by Dyens as well
as their resulting tendency to signify multidirectionally at the very moment that they are intended to be most unilateral. In this respect, Dyens
cannot adequately answer accusations of "slapdash science" because his
construction of the cultural body is ultimately dependent on this slippage.

Indeed, complaints of unscrupulousness are precisely what regularly
met *Metal and Flesh*'s publication, especially in the scientific community.
For example, Philip E. Mirowski is perplexed as to how anyone could "so
lumpenly reify 'memes' to the point that they might convince themselves
that something called a 'cognitive ecology' could actually displace the biological ecosphere."[66] Furthermore, he accuses Dyens of using "watered-
down summaries" in lieu of "real" scientific sources, resulting in numerous
theoretical errors that ultimately amount to a book that is a "projection
of naturalistic metaphors onto cultural studies."[67]

What this criticism fails to recognize, though, is that the metaphors
slide both ways and that "just because a particular discourse operates
within parameters and conventions that we think of as 'scientific,' . . . [this]
does not mean that the discourse is not metaphysical."[68] That is, there is
a sense in which Dyens—perhaps naively—has simply taken science at its
word: if scientific knowledge can truly be measured, verified, and added
to an existing body of disciplinary knowledge, then what is to stop that
knowledge from being radically decontextualized in the way that Dyens
has done? The answer is certainly not disciplinary boundaries, because

science—as a whole—alleges to describe material reality (rather than
the materiality of its discourse), which is also Dyens's task. Moreover,
even specialist knowledge cannot be convincingly invoked: Dawkins, for
example, explicitly states the folly of making a "clear separation between
science and its 'popularization'" because pushing the novelty of language
and metaphor can produce a new way of seeing that in its own right makes
"an original contribution to science."[69] Moreover, as N. Katherine Hayles
succinctly states, "metaphor is not opposed to scientific work but intrinsic
to it."[70] In effect, then, Dyens simply accepts the implicit claim that because
evolutionary theory has been scientifically verified, it describes a material
process that is fundamentally and extradiscursively true and that therefore
obtains equally in any discursive setting. If *Metal and Flesh* is objection-
able to the scientific community, then this may be because Dyens accepts
that scientific results are data in the true sense, namely, that they describe
something that persists through a process of extraction and reinsertion
(which is to say, they are quantifiable).

Consider the alternative: if scientific data—and thereby scientific
knowledge—were context dependent to the extent that their representation in language was *constitutively* inadequate—to the extent, that is, that
their results could never be rendered as data that could be abstracted from
their local circumstance—then by what means would this knowledge be
verifiable? Indeed, what would constitute this as knowledge?[71] Certainly
one can make a claim for moderation (i.e., that data might be abstracted
but only to a certain extent), but as long as this claim is delineated in terms
of disciplinary boundaries, it is destined to limit any findings to the role of
recapitulating (and naturalizing) the logic of previous findings within that
discipline. The point is that while principles of scientific reality (and scientific exchange) are undeniable, the science itself—taken overall—cannot
be exchanged for anything and is thus unintelligible to that which it does
not determine. In the case of evolution, this is the (a-rational) scientific
worldview from which it springs.[72]

Thus the fact that Mirowski—among others—objects so strongly
to Dyens's work might instead suggest an unspoken desire to reject the
latter scenario: to claim its constitutive neutrality, the scientific method
needs to be able to act as though its discoveries are objectively true, an
ideological comportment that is thus built directly into the very claims

that it evinces.[73] That is, a particular method might be modified or refined to produce data that contradict earlier findings, but the possibility of producing data is the grounding principle of scientific inquiry. This tenet is spurious because it allows any particular instance of method to transcend the specific differences between it and other methods, thereby borrowing an unacknowledged value claim of method itself to lend authority to the particular instance. Simply put, science necessarily assumes that there is a method. Mirowski's conscious objection to Dyens's liberal use of scientific metaphors, then, may bespeak a deeper anxiety about the truth value of the referents of the metaphors themselves, an intuition that "science can no longer comprehend itself as a representation of the world as it is, and must therefore retract its claim of instructing others about the world."[74]

So Dyens has made scientific errors, yes. However, his argument would have been incoherent had these errors not been made because the errors mark an (unconscious) attempt to reconcile the incommensurability that sustains the metaphysics of the evolutionary science that grounds his argument. In this, Dyens illuminates the gap between Dawkins's selfish gene theory's (and, more broadly, science's) idea of itself and its reality, namely, its partiality with respect to motivation. In so doing, he approaches the "why" of science, asking why it values what it does. It may be true that Dyens's understanding of science is flawed, but his proposition—as an actual utterance—needs to be deciphered.

Indeed, the ramifications of this deconstruction are not only deleterious to the scientific objectivity that underwrites Darwinian evolution. Instead, the desublimating movement flows back toward Dyens, whose flaunting of scholarly convention sometimes teeters on the edge of simple unscholarliness in its numerous contradictory claims that are left unpacked. In this respect, *Metal and Flesh* additionally merits study for the position that it occupies with respect to other scholarly research: the text carries the intellectual clout of being included in the Leonardo series published by the MIT Press, and yet it is a short and eclectic study that exhibits little of the scientific weightiness that characterizes that institution.[75] Furthermore, although the book certainly stirred controversy when it was released, it is safe to say that it has not led to Dyens's universal acceptance as a major thinker of the convergence of art, science, and technology (the purported area of the Leonardo series). Might *Metal and Flesh,* then, be considered a footnote to the broader scientific text of MIT Press? If so, this status would suggest the text as a promising site to begin a deconstruction of the values implicit in the writing of scientific posthumanism: what unmarked privileges, for example, might be mobilized by the very gesture that seeks to uncover the laws that govern replication? Dyens answers this question—performatively—by flipping the scientific narrative of Darwinian evolution to write an evolutionary narrative of science.

IN A RECENT LECTURE, Dyens illustrated the materiality of cultural bodies by asking a simple question: why would one make clones when one has advertising?[76] That is, why be constrained by an outdated notion of biological materiality when all the necessary resources exist to create *cultural* clones? The implication of this is double: first, we are controlled to a greater extent than we can imagine by our cultural (including biological) situations so that (from a behavioral perspective) the science of cloning is a fait accompli; second, though, this also implies that clones themselves are no longer "elsewhere" and "other" but instead have come to occupy a legitimate place in our existential universe. If the behavioral claims that Dyens makes are true, then—in all probability—contemporary subjectivity is overdetermined.[77]

However, if this projection of scientific posthumanism suggests that such overdetermination is specific to our present era, the constitutive (often implicit) claims contained in scientific posthumanism signal the reverse: the subjectivity of scientific posthumanism is a mythic one in McLuhan's sense[78] in that it is the static abstraction of representations from the living process of evolution. As such, it can be understood as an "intelligible explanation of great tracts of time and of the experience of many processes, and [can be] used as a means of perpetuating such bias and preferences as [it codifies] in [its] structure."[79] Captured under the sign of a single logic (that of Darwinian evolution), the hyperspeed of fractal subjectivity retrieves the deep time of myth: the celebratory hybridity of scientific posthumanism belies an abiding loyalty to the ideologies that produce it, ideologies that date back (in their present form) to at least the Enlightenment.

Thus scientific posthumanism not only retrieves its avowed content (namely, the primordial soup of genetics, which it plays out in the scene of cultural bodies) but also death in its fullest sense. That is, if we accept Dy-ens's argument, then death no longer simply indicates the nonliving (which cultural bodies absorb into a continuum of life) but also suggests a form that persists even after death has been obsolesced: technological culture's evasion of ontic death renders death an important ontology of our time in that it is retrieved as a total and inevitable (and thus noncontemporary) fate. Simply put, Dyens offers deep-seated nihilism as a foundational myth of (both genetic and memetic) Darwinian evolution—not just the end of the human that he predicts but also an end that he performs.

Indeed, if Dyens narrates a detachment of culture from the previously dominant genetics to reproduce along its own memetic lines, his text itself points to these cultural bodies ultimately detaching from culture in a generalized diffusion of negative replicators: the memes for not being human, for not being embodied, for not being culturally situated, and for not having agency all replicate with astonishing velocity.[80] When life is all-inclusive, the negative genes that we all possess are innumerable, so who is to say how they will interact, splice, and replicate, and even what replication would mean in this context? In short, cultural bodies are nihil-istic because they are dominated by everything that they are not and are thus overdetermined by the nihilism of their subjection.

In this context, it bears noting (vis-à-vis Baudrillard) that the vitality of life lies in the clear opposition between life and death, whereas death—in the full sense—lies in the lack of distinction between the two. That is, when there ceases to be a substantive difference between life and death, death takes on its full ontological implications—its fate—precisely in the form of life.[81] If technoculture—in Dyens's reading—disavows death, then, this does not mean that it has eliminated it but rather that it has expelled it from its system of values.[82] As a result, we may well be approaching Ray Kurzweil's "singularity," but his dream of longevity may turn out to be the ultimate dystopia, in the sense that the metric of life will have irrevocably changed.[83]

In summary, the scientific posthuman is thus a posthuman divested of subjectivity and trapped in a language of scientific value, where it survives autovampirically. To be clear, this is not an argument against the logic of evolution, or even against the logic of cultural bodies: in-deed, one of the ironies of contemporary technoculture is that it reveals compelling evidence that the boundaries between humans and nonhumans, including animals, are more porous than we might think.[84] Rather than contesting these facts, then, I am simply asking what the facts *are*, in the sense of what they mean. The answer to this question is not—and cannot be—an empirical observation or definitive categorization; instead, it consists in the insistence that the entrance of these observed facts into discourse is also the moment of their constitution as facts, an amalgamating-amalgamation in which cultural bodies begin to speak the foundational biases of science.

2 DARK MATTERS: AN *EIDOLIC* COLLISION OF SOUND AND VISION

One can look at seeing; / one can't hear hearing.

MARCEL DUCHAMP, CITED BY KAHN IN *NOISE, WATER, MEAT*

It was terrible, . . . with breath which one could almost see rather than hear.

ITALO CALVINO, *THE BARON IN THE TREES*

To the extent that scientific data are repeatable, verifiable, and falsifiable, we might follow McLuhan in saying that they are staged in visual space.[1] In chapter 1's analysis of Dyens's cultural bodies, then, it is specifically the *vision* of scientific representation—which is virtually synonymous with replication, for Dyens—that is challenged from two sides simultaneously: on one hand, it is haunted by forces that are qualitatively different than those that it registers; on the other hand, its very representations, pushed to the extreme, turn back on themselves to speak against the terms that conditioned them in the first place. In a sense, these two lines of encounter subtend one another, the former taking place as that which is present but invisible to empiricism—what I will call its *sound*—and the latter as that which empiricism creates but cannot avow as a creation.

If scientific technological posthumanism (of the type emblematized by Dyens) is situated in visual space, then, it is not difficult to understand how it might be haunted by sound: like Serres's parasite, sound plays the position, while science—conceived and constrained by visual objectivity—tends to play the contents.[2] In this sense, every sound is a ghost, which is why we are always looking for the sources of sounds, trying to place and identify them: we hunt down hauntings and flush them out, only to always hear more. Indeed, when all is quiet and calm, we still hear blood circulate through our veins with a gentle chatter and thump that suggests that we ourselves are ghosts: to be alive in an age of cultural bodies is to be haunted, be it by genes, memes, or any other strata of reproductive otherness.

However, how might we think the other intensity of scientific technological posthumanism, that which is registered visually, created in visual
space, and yet—in the very fact of its being created—speaks to a force that
is invisible? In fact, scientific discourse has already nominated this phenomenon—or rather, this nonphenomenon—as "dark matter," marking
an undetectable hypothetical substance that structures the entire universe.
Eluding the ubiquitous pull of electromagnetic force, dark matter is visible only through its invisibility, through its gravitational pull on visibility
itself. Darkness, it seems, is what we are really perceiving when we see
the spectacular explosions of charged atoms that constitutes, equally, the
atrocities of Guantanamo and the click of a mouse, the underground activity of the Large Hadron Collider and its full spectrum realization in the
language of digital media. History itself, dark matter would suggest, can
only be a phantasmatic relation to the real gravitational pull of darkness.

Following the lineaments of these two related but distinct metaphors—ghosts and dark matter—this chapter probes the exhibition *Eidola,*
which unfolds an encounter between disciplinary biases of sonic and visual
art practices that accentuates how both are infused with a part of the
other that they simply cannot make sense of.[3] To this end, the exhibit pre-
miered works by two artists: *False Ruminations* by William Brent, a sound
artist and computer programmer whose works are regularly realized by
robots that possess a striking visual presence, and *Basement Suite* by Ellen
Moffat, a visual artist whose installations employ sound environmentally,
compositionally, and as a catalyst for audience immersion. In this chapter,
then, I will argue that *Eidola* operates in and as an incommensurability
of sound and vision, which the exhibit works against one another in a
riot of recombinant and reiterative hauntings and forces. In this reading,
sound intervenes as a dual identity, on one hand whispering its otherness to vision, while on the other hand suggesting that this otherness is
dominantly interior to vision as the "paradoxical identity of both sides
of the distinction."[4] In this way, sound—like dark matter—is a blind spot
of visual observation but as such operates equally to trouble both terms
in the relation as it does to assert the relation itself. Dark visions become
haunting sounds and haunted sounds are darkly visible, then, with each
flip contributing to a collective insistence on a collision of sound and vision that, with *Eidola,* spawns the question: in hearing (rather than seeing)

dark matter, might it be perceived not as a structuring principle but rather as a kind of sonic delirium?

ENTERING THE GALLERY for the opening of *Eidola*, your vision circulates in series between three locations: upward to Brent's "Ludbots," sixteen inverted percussion mallets housed in industrial sheet metal bases and suspended from the gallery ceiling in a disciplined stillness, all glaring at the lengths of wood that hang inches from their heads; across to the deepest corner, where the skeleton of a horizontal surface floats above the gallery floor, a single bottom-mounted ethereal light acting as an arterial link between the network of widely spaced floor panels; and finally, around the room, in search of familiar faces, social cues, or simply a place to stand. Each of these three figures draws only a moment's attention before your eye moves on to the next or back to the previous; the order of movement is unpredictable but relentless and motivated.

Supplementing this vision, though, is the sonic component of *Eidola*, which is, in a sense, perfectly wrong for an art opening setting: in *Basement Suite*, the slow and quiet revelation of the everyday melody of a house creaking, furniture dragging, and children scampering—all projected through small speakers mounted to the bottom of the suspended panels—is lost amid the throng of conversation and the squeaks and groans of the gallery; conversely, *False Ruminations* alternates between sitting in quiet stillness and bursting out in exuberant four-minute attacks of mallets on wood, the latter aggressively interrupting the social gathering of the opening to insist on themselves. The patterns that emerge from this insistence, though, are not supported by the accoutrements of the space: the rituals and spacing for conventional listening are simply not available so that the would-be sonic phrases operate instead rhetorically, as belligerent interjections, rather than as a site of attention in their own right. Collectively, then, the sound of the installation is at best inappropriate for the setting of the opening, being alternately inaudibly soft and inconveniently loud.

In this first encounter with *Eidola*, then, two ghosts emerge, each paradoxical in nature. In *False Ruminations*, the sound is clearly audible but is rendered visually: the literal sound is alternately present and absent and functions unambiguously and autonomously. This clear figuration, though, disinvests the work of its sound per se—of its sonic mediality—in

FIGURE 1. Ellen Moffat, *Basement Suite* (installation detail from *Eidola*, Open Space Artist-Run Centre, Canada), 2009. Photograph by Ted Hiebert.

FIGURE 2. Exhibition shot, *Eidola*, featuring work by William Brent and Ellen Moffat, curated by David Cecchetto and Ted Hiebert (Open Space Artist-Run Centre, Canada), 2009. Photograph by Ted Hiebert.

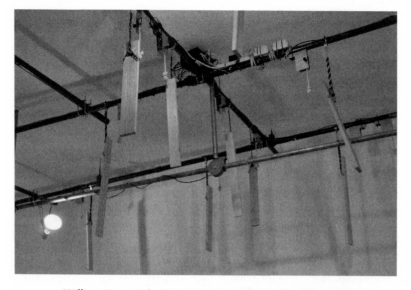

FIGURE 3. William Brent, *False Ruminations* (installation detail from *Eidola*, Open Space Artist-Run Centre, Victoria, Canada), 2009. Photograph by Ted Hiebert.

the sense that these interruptions are (at the exhibit's opening) haunted by the absence of any specifically sonic relation. That is, if we are thinking of sound as a kind of dark matter, the fact that *False Ruminations*'s sounds are identifiable obscures the darkness—the unvisibility—of its material. Put differently, *False Ruminations* makes the "sound-ness" of its sounds disappear in full aural view, their sonic specificity sublimated to their registration as both the sound of the Ludbots and the presence of Brent's piece (which did not, prior to the Ludbots' activation, otherwise separate itself from the collective exhibition). That is, these interjections are medially underdetermined: they operate functionally, and their function is one that could be performed without, necessarily, any sound at all. In other words, their sound is put into discourse.

By contrast, *Basement Suite* works in the opposite manner. The time between *False Ruminations*'s eruptions—each interval lasting as long as eight minutes—would, in theory, be the time during which the comparatively soft sounds of *Basement Suite* are exposed. However, the opening's large attendance renders the sounds from the *Suite* indiscernible from those of the gathering: the piece is by no means silent, but it is also not identifiable.

And yet there is always the sense that the clusters of people are grouped in specific formations, that their social chatter falls subtly in rhythm with the phrasing of Moffat's environment; indeed, the composition of *Basement Suite* consists in recordings made with microphones attached to the underside of her home's floorboards, above which social gatherings much like the opening took place during recording. The point, then, is that *Basement Suite*'s sounds are practically inaudible, but as a result, in hearing them, we lose ourselves to their ghostly sonic physics.

The conversation at the opening, the social gathering, is thus ghosted in its own right, haunted both by a present absence (the flight of medial specificity from *False Ruminations*'s sounds) and an absent presence (the social organizing effect that *Basement Suite* has on the attendees). These first ghosts of *Eidola,* then, are the twin movements of sound from and into perception, each instituting a degree of disjunction into the opening that is congruent with the overall thematic play of *Eidola*: the title of the exhibition, after all, was chosen for its double etymology, tracing its lineage to a Greek word that means both "ghost" and "bias." Between these three scenes that first catch one's eye on entering, then, a cross-pollination occurs as sounds emerge through object and social presences. And yet the reverse is also true, in that objects and conversation emerge as the material detritus of sound. Sounds are channeled into spatial dialogue, ghosts sing with ghosts through bodies of wire and wood: an interdisciplinarity of new media practice, realized through practices of disciplinary ghosting. *Eidola* asks, in other words, what happens when sounds take visual presence? No longer dematerialized—no longer simply auditory—sounds are here seen as well. Such is the demand, and the demand is no less for the object that insists on being heard. No longer passive visual presence, these objects speak in tongues through the audience that attends them.

OF COURSE, the exhibit is not designed with the opening exclusively in mind, and entering the space at any other time gives an immediately different impression. No longer a social intervention during these quieter times, the sense of *Eidola* as an exhibit gives way to a visceral realization that it is an installation. Simply put, it is immersive, so that one feels less inclined to marvel at the play of presence and absence than to investigate the complex patterns that take place over time. In particular, these patterns emerge from a doubled bias of sound and vision, where, paradoxically, to see something is already to have answered the question of the origins of sound, to have sighted the ghost. And yet *Eidola* suggests that this is the point where imagination flips into hallucination, where we begin to perceive that which is not there, even if it used to be. That is, the most marmoreal visions of *Eidola*—the still Ludbots, the architectural panels, the looming space of the gallery itself—are the most unsettling because we lose our purchase on the line separating the real and the imagined, between the object and its emanations in the myriad sonic tendrils that escape from it. Similarly, there is a sense in which *Eidola's* sounds are always heard, as sounds, before they are sounded, before they become the sound *of* something. That is, in *Eidola,* we hear a sound by losing ourselves—momentarily—to its ghostly physics: the moment of hearing is one of jarring disjunction and, conversely, the moment that a sound is registered is also the moment that it loses its sound-ness to the collapsed dimensions of our presence. In both its sounds and visions, then, the gambit of *Eidola* is not an irenic conflation of the senses but a struggle between alternate realities that cannot both, sensibly, be true.

This is precisely what is at stake in *False Ruminations,* where one intuitively feels the difference between sound and sounds, between an unknowability of sound itself and the sound-objects that contemn this mystery and seek to contain it. We feel this first when we look at the Ludbots, robot instruments that are strange only because they have left their percussionists behind and absconded to the rafters of the gallery. Even if one first spots them while they are suspended in stillness, inverted, there remains the sense that their objective presence is foremost an alibi for an event to come. We hear—with our eyes—a world of possible sounds, even as the work itself is "silent."

This potentiality is played out in the musical composition of the work, too. Prior to the installation, an initial rhythmic sequence stole from Brent's hands into the computer, where it quickly propagated into the roughly fifteen thousand sequences that now scurry about the wires. Revolving around a custom algorithm, then, each iteration of *False Ruminations* is an execution of two transitional processes that take place in parallel: two of these offspring variations are selected as starting points (A, C), and two are selected as ending points (B, D); each performance of the piece is simply the simultaneous convolution of A into B and C into D. The point, then, is that the Ludbots themselves—that is, the robots—exist first as a kind of material detritus of *False Ruminations*'s sound. They are not instruments per se because they are not instrumental to anything. Instead, they are what is left after the digital commingling has ceased and the algorithmic dust has settled.

This, in fact, is why the Ludbots are ghostly: the distance between them and their performance—the distance between the composition and its interpretation—is closed, because the algorithmic construction of the composition means that it literally does not preexist its performance. That is, the specific contents of A, B, C, and D are determined only in and by the piece's performance. In turn, this content informs the transitional processes because the latter are calculated as convolved signals of the start and end points (i.e., A and B). In this, then, a paradoxical (non)relation is articulated between the composition and its performance, echoing Serres's assertion that if a relation is perfectly immediate, "it disappears as a relation [such that] relation is non-relation [and] the real is not rational."[5] That is, we move with the Ludbots from listening to them "play a piece" to simply listening in a place where we no longer have the distance from the piece to stand back and see who is manipulating it as such. Here we are indeed in a ghostly territory where we hear everything—patterns, timbral and spatial groupings, transitions, and progressions—but also know that everything, in a sense, isn't.

If Brent's *False Ruminations* can be so seductive in its aggressive yet rhythmic variations, is this not also because it makes us wait during its periods of nonactivity and, in so doing, incorporates us, too, into its algorithmic process? Ghosted by the very work we witness, the viewer–listener is part of the mathematics of the equation, even if for no other reason than to give us the opportunity to wander into *Basement Suite* undistracted. And as we do so, the difference between Brent's and Moffat's work is marked: whereas *False Ruminations* sounds an irrational machinic agency through a temporal performance, *Basement Suite* reveals itself in an objective space of architecture: the small platforms in the piece are hung schematically, referring back to the original layout of Moffat's own basement. Indeed, ethereal lighting casts shadows on the gallery floors that support the material reference to this other basement architecture with a mythic dimension, working in tandem with the sounds of plaintive flooring to suggest the mysterious foreboding that has long characterized cellars. These salvaged icons of the floating floor, fragmented into only those platforms that matter, ground the underground echoes of times arranged and reanimated. Yet, at the same time that the iconography of a basement anchors the platforms from below, they are physically suspended from above. What we are left inhabiting, then, is neither the basement, nor the floor above it, but rather the boundary between the two, the place where a cozy kitchen becomes a haunted cellar. Indeed, the de facto viewing position supports this liminality, as we stand with our eyes and arms above the flooring panels, but with our feet planted firmly below. In a sense, then, the architecture of *Basement Suite* mediates an encounter within ourselves.[6]

However, to the extent that the physical presence of Moffat's piece disambiguates these two domestic settings, its aural component works in the opposite direction. That is, we expect skeletons and ghosts and mysterious sounds of the past to come from the basement—to come from below—but what we hear beneath us in *Basement Suite* are sounds from above, sounds of high heels and dancing and tuning forks and dinner parties and the dragging of furniture across the first-level floor. Or rather, we hear these sounds of above from below *and* above: the speakers through which they are sounded are mounted to the bottom of the platforms, but the platforms are positioned at a height that allows us the option of listening from different vertical vantage points. In this, the platforms themselves are both the basement floor and the basement ceiling so that we inhabit the constitutive ambivalence of this boundary, positioned upside down and right side up at the same time.

If *Basement Suite*'s sound works in opposition to its physical presence, though, the reverse is also true: because of the simplicity of form in the work's appearance, a certain vertigo of disappearance takes place. That is, the ambiguity of Moffat's aural space is only really remarkable for the visions it invokes: ghosts speaking to, dancing with, and rearranging furniture with other ghosts. As a result, *Basement Suite* shows its logic of the living to be contaminated with "the distinct and always concrete operation of technics,"[7] but in an ambivalent way: it is not simply that the house—or the basement—is haunted. Instead, because the piece is immersive, it is we who are doing the haunting (even if we are only haunting ourselves).

Moreover, our haunting is also haunted, for who is this haunting "we" if not subjects who are ghostly present?

THUS EIDOLA explores the incommensurability of sound and vision both within and between the individual works. However, the preponderance of reversals that emerge suggests that this antagonism is not the whole story. Indeed, beyond a co-implication of visual and aural ontologies, Brent's and Moffat's works step in to insist on themselves in their own right. The sense of productive antagonism remains, but so, too, does a serendipity that can't be ignored: as the artists worked to complete their respective works for the exhibition, they were each compelled—without any knowledge of the other—to dramatically change their construction materials. What they arrived at independently were pieces with striking similarities ranging from their sharing a sparse visual aesthetic to their both making prominent use of wood and light industrial materials. Considering that the only curatorial directions given during the production stage were aimed at emphasizing the disciplinarily informed differences between the artists, this confluence is truly remarkable.

In fact, these similarities ultimately play an important role in the exhibition as a whole. Emphatically, they do not suture the gaps that *Eidola* unearths but rather emphasize the ambivalence with which these gaps signify. Thus, for example, if the teleological orientation of *False Ruminations*—the fact that each iteration moves through a predefined, audible process—serves to orient the piece as an atemporal object in space, this very "object-ness" is amplified by *Basement Suite* to paradoxically point to the indeterminate temporal framing within which the piece exists—indeterminate because it hinges on the embodied—and thus singular—activities of the listeners who move with the piece. Similarly, though, *Basement Suite* certainly articulates an architectural space that can be perceived objectively, from different angles; and yet, doesn't a melody of sorts begin to emerge in it—not only in its creeks, crackles, and complaints but also in the panels that lie lithely still superjacent to these sounds, a strain that is all the more tuneful for the abrasive intrusions of *False Ruminations*?

What *Eidola* makes, then, is a claim that sound might act as dark matter in the exhibition; it might act, that is, as a kind of presensory organizing principle from which the relations that make sense of its collisions of sound and vision emerge. If this is the case, though, *Eidola* also suggests that its sound—like relation for Serres—is also nonsound, just as dark matter is also peripherally visible (in that its effects are seen when we look elsewhere). In this latter formulation, then, a tautological bias is installed in the exhibit because it justifies its claims—necessarily—through the same logic that the claims themselves are made. Simply put, if sound is the ground of *Eidola*'s sound–vision relation, this status is itself dependent on a notion of both sound and vision that preexists the exhibit. That is, the sound of *Eidola* is haunted by both sound and vision, by an unavoidable ambivalence that it stands in for.

If *Eidola* is predicated on a sense of disjunction between sound and vision, then, this disconnection is felt through a kind of sixth sense that intermingles these strata (even if it does not literally connect them) without being determined by them. This, then, is the mixed reality of sound and vision that *Eidola* institutes, a dark matter that congeals these opposed senses into a sound exhibit (i.e., into *Eidola*) that somehow makes sense: we hear *Basement Suite* as much with our eyes as our ears, just as the temporal algorithmic logic of *False Ruminations* is seen, with our ears, as an identifiable unit (that is, as a "piece," in the musical sense). The exhibition's oppositions each flow into one another, then, at the same speed that they fly apart through the works' similarity to one another. The point, ultimately, is that *Eidola*'s sound and vision do not reproduce in the way of Dyens's cultural bodies, memetically transferring their own (non)genetic materials. Instead, they resonate with one another in time to a ghostly trajectory that is not contained in either of them, individually.

IN THE SAME SENSE that *Eidola* performs a dark matter of relation (i.e., a presensory sound), it is a kind of dark matter that Hayles is pointing to in the short but devastating critique that she levels against Dawkins's *The Selfish Gene*: unpacking the slippery rhetoric of Dawkins's argument, Hayles shows how Dawkins sneaks a "motive force" into his argument by undermining the importance of his characterization of the gene as a protagonist (he puts it down to his own "sloppy language"). In contrast, Hayles demonstrates that this force is necessary for Dawkins's argument to function causally, rather than being a simple description of "shifts in populations that can be statistically measured."[8] This shift, then, combines with Dawkins's radical decontextualization of genetic processes[9] to allow him to solidify his own agency "even as he supposedly gives it away."[10] Thus, though Dawkins is completely invisible within the selfish gene narrative, Hayles notes that he is nonetheless located—or rather, that a particular construction of Dawkins is located—in the authorial pull that he exerts on it.[11]

What *Eidola* says about dark matter, though, is slightly different than what is evidenced in Hayles's argument with Dawkins, if not necessarily opposed to it. In *Eidola,* the sound of dark matter undermines the cultural bodies of sound and vision (i.e., undermines their representations of materials as sonic or visual), but this sound is itself haunted. In short, *Eidola* stages an argument with dark matter, an argument that ends with the insistence that just because the terms (i.e., sound and vision) are questionable, that doesn't mean that we ought not to use them. Neither, though, does the necessity to use them mean that we ought not perpetually interrogate the exclusions by which they proceed. In *Eidola's* dark matter, then, sound and vision are ghosted into their full ambivalence: on one hand, the relationality of sound is the outside of vision, that which is excluded from visual identity, that which exceeds the speed of light by never casting a shadow, the "invisible . . . [that] remains *heterogeneous* to the visible";[12] on the other hand, the relationality of sound is included in vision as its constitutive outside, as that which is necessary for identity, as that which has no speed because it has no positive-substantial materiality. As such, we can also hear the pull of this dark matter—the pull of sound itself, performed in a theater of vision—as a push, we can hear the movement to relativize vision to sound as the double-move to consolidate sound itself, a relational nonrelation, in the language of vision.

Ghosting Judith Butler, we might note that the magic in every good magic trick is worked in its setup: if the magic of science is that it allows us not only to see in the dark but to see the dark itself, the trick lies in the fact that the darkness that science addresses was never really dark but only invisible.[13] Darkness isn't illuminated by science, which would be to make it contingently visible and thus to feed this contingency back onto reality itself. Instead, darkness is robbed of its darkness by science precisely because it is rendered *in*visible, because it is always dark *matter;* the constitutive ahistoricity of the language of the universe is always citationally and historically constituted through the language of history, which is in our time the language of science. The term *dark matter*, then, is the constative claim that it isn't "all relative" after all, a constative claim that simultaneously performs the insistence that it is all relative, but radically so.

Thus what Dyens finds in Dawkins is not just bias built into science but a point where science is haunted by the necessity of having such uninterrogated presuppositions. Indeed, this observation is what is performed in his intervention and marks both its merit and limit: in *Metal and Flesh*, a discourse of Darwinian evolution is rendered as a discourse, which is to say it is reduced to the descriptions that attest to its truth. In this, the discourse becomes weightless: it is no longer anything but a reproductive performance, exchanging in itself rather than life and circulating without reference or circumference.

This same hyperrelativity—outside the language of physics—takes place in a different way in *Eidola*, which performs a productive antagonism of sound and vision that speaks in ghostly tongues of an orientation that underwrites them both. In this sense, *Eidola* is ultimately an installation of the psyche, in all its genetically and semiotically underdetermined wonder: it doesn't so much institute new technological realities as it opens the door to a reality of hallucination and imagination. In this context, the disjunctions that constitute *Eidola's* rhetoric are reframed by its performance: disjunction doesn't mean anything, doesn't have a negative value, when instruments play themselves with slow-rhythm syntax and neural networks speak in fractured tongues, strangely singing along. Here ghosts grow voices of their own that emphasize the connections between automated voice, sound, and presence. But in this emphasis, paradoxically, it is precisely the disappearances that emerge, front and center. These disappearances are confrontational because they won't go away: they are hauntings but also real voices that are reproduced in phantom spaces; they are ghosts in the machines that also ghost those that surround them, implicating their very audience in the witnessing of impossibility. After all, what are ghosts if not autonomic consciousness—automated intentionalities that infiltrate the gallery space—daring us to insist that we are different?[14]

Ultimately, then, every ghost is a sound too, a lingering heartbeat that came from somewhere and somehow strangely persists—persists perhaps even precisely because its strangeness refuses to be reincorporated back into a world of forces, pulls, and identity. *Eidola*'s sounds have no sensible material presence: they separate from their origins to travel—out of body—until they collide into *us* and reintegrate. Ghosts from the past, they haunt us with their relentless chatter, even while their voices dissipate into the world, the speed of their fading exactly equal to a speed of sound that, because it is slower than light, grounds the latter.

PART II

N. KATHERINE HAYLES AND HUMANIST TECHNOLOGICAL POSTHUMANISM

Knowledge is always *for*... some things and not others.

DONNA HARAWAY, "MORPHING IN THE ORDER"

In this chapter, I discuss the humanist technological posthumanism that is proffered in the writing of N. Katherine Hayles.[1] The chapter begins by situating Hayles as a coalescing figure of posthumanism, discussing her role in forming a coherent discourse around the various and scattered activities that have slowly combined to overdetermine the cybernetic landscape. In this perspective, Hayles's insistence on the historical specificity of this discourse is considered, with special attention given to the new constructions of materiality that emerge with contemporary technologies. In particular, the chapter limns Hayles's considerations of code, narrative, and language as well as her adhesion of a dialectic of pattern and randomness to the existing subjective dialectic of presence and absence: without making explicit reference to sound (in the sense that the term is mobilized in this book), Hayles's project is intimately engaged with concrete relationality and with the intermediating feedback loops of disparate media. Thus, from the emphasis on embodiment that is entwined with her perspective, Hayles inscribes an ethical dimension into her posthumanism.

This inscription forms the most significant site of critique offered in the chapter, which concludes by arguing that Hayles's construction of technological posthumanism ultimately reinscribes the humanist ethics that it purportedly moves against. Moreover, precisely because Hayles stages this inscription as a *post*-humanist one, its humanist grounding is sublimated in an ironic reversal of posthumanism's basic tenet (i.e., that the human is a subject of discourse). Thus, while the politics that accompanies Hayles's ethical comportment may be admirable, I argue that it is ultimately bound by many of the same constraints that plague the very humanist discourse she denies.

WHILE MANY well-known humanities theorists (including Derrida, Lacan, Deleuze, Foucault, and Haraway) have explicitly contributed to the discourse of posthumanism, the term itself only began to take on real currency in the humanities (especially in North America) with the publication of Hayles's *How We Became Posthuman* in 1999. With this book, Hayles solidified the posthuman as a meaningful way to discuss how "a historically specific construction called the human is giving way [to something else],"[2] while simultaneously opening the term to the myriad instantiations that have come to characterize its use today. In relation to Hayles, these proliferations are both integral to her thought and mobilized against it: the former in the sense that she consistently emphasizes the importance of understanding meaning-formation as a continual process, and the latter in the sense that significant critical energy has been spent contesting many of her claims and crafting alternative metaphors to those that she has proposed. All told, then, Hayles is perhaps *the* catalyzing figure of technological posthumanism as it is found in the (digital) humanities, even when the discourse moves against her.

The irony of Hayles's occupation of this position is that *How We Became Posthuman* is framed as a critical intervention into an historical discourse that had already been formed. Specifically, the text seeks to unpack "how information lost its body [and] how the cyborg was created as a technological artefact and cultural icon."[3] We might observe, then, that *How We Became Posthuman* retroactively constructs a unified narrative of technological posthumanism that it simultaneously multiplies. Put differently, Hayles utilizes the humanities' powers of historical narration to demonstrate the disavowed choices that had served to naturalize the (at that point still more or less unnamed) posthuman as it was constructed by the discourses of cybernetics. Her nomination of the term itself thus participates with these sublimated decisions to constitute the type of feedback loop that is frequently found in her texts, with *How We Became Posthuman* produced by the very history that it constitutes as such (rather than as a series of discrete facts). In Brian Massumi's terminology, Hayles "retroduces" posthumanism.[4]

Toward this end, the point that Hayles makes more forcefully (and more often) than any other in *How We Became Posthuman* is that we must understand information to be dynamically entwined with embodiment.[5]

That is, neither embodiment nor information (nor thought, for that matter) precedes the other; instead, the two are coextensive (though conceptually distinct). As such, Hayles offers Hans Moravec's dream of downloading consciousness into a computer (among other examples) as emblematizing a conception of information that is fundamentally ideological: a reified concept of information is treated by Moravec, she argues, "as if it were fully commensurate with the complexities of human thought."[6] By contrast, Hayles does not see embodiment as a realization of information but rather as the being of the data itself.[7]

Importantly, Hayles notes that the dream of erasing embodiment—of which Moravec is only one inciter—is not unique to cybernetic discourse but is also a feature of liberal humanist subjectivity. Thus, for example, she remarks that it is only because the liberal humanist body is not identified with the self that it is "possible [for it to claim] its notorious universality, a claim that depends on erasing markers of bodily difference, including sex, race, and ethnicity."[8] In this respect, then, the version of posthumanism that Hayles seeks to unseat represents a continuation of liberal humanist subjectivity rather than a break from it (and thus would likely today be termed "transhumanism"). In contrast, Hayles hopes to "replace this teleology of disembodiment with historically contingent stories about contests between competing factions."[9]

In accounting for the prevalence of the disembodied posthumanism that she contests, Hayles identifies three distinct waves of cybernetic research. She names the first of these waves the "Macy period," after the Macy Conferences on Cybernetics that were held from 1943 to 1954, and situates the period as roughly spanning from 1945 to 1960. The key thinkers of this period were the participants of this conference, including such diverse luminaries as Norbert Wiener, John von Neumann, Claude Shannon, and Warren McCulloch.[10] Though Hayles emphasizes the important differences between these thinkers, she also notes that they are joined by their shared interest in homeostasis. In particular, thinkers during this phase of development did not so much seek to enact specific stable actions through machines and cybernetics as to understand the conditions under which stability itself could be achieved. Put differently, these thinkers were responsible for shifting from the existing emphasis on finding solutions (characteristic of robotics) to the cybernetic goal of stating problems.[11]

As Hayles suggests, thinking of information as a signal–noise distinction is intimately tied up with this new accentuation but also dematerializes information and divorces it from questions of meaning. The second wave of cybernetics identified this, stated it as a problem, and attempted to offer a corrective through its emphasis on the radical epistemology of autopoietic reflexive systems.[12] Spearheaded by Humberto Maturana (and stretching from 1960 to 1980), work in this period is predicated on understanding perceptive apparatuses as not so much *representing* reality as *constructing* it. To this end, Maturana's powerful critique of objectivist epistemology concludes that although reality exists, it comes into existence *for a particular organism* (including humans) "only through processes determined solely by the [perceiving] organism's own organization."[13] Because a given system thus constructs its environment through the "domain of interactions" made possible by its autopoietic organization, "what lies outside that domain does not exist for that system."[14] As a result, "the space defined by an autopoietic system . . . cannot be described by using dimensions that define another space," but when we refer to our *interactions* with a concrete autopoietic system, "we project this system on the space of our manipulations and make a description of this projection."[15] As such, this period is characterized for Hayles by its reinvigoration of the specificity and concreteness of embodied processes.

Furthering her account of this "reflexive" stage of cybernetics, Hayles discusses some of the ways in which Maturana's concept of autopoiesis was further elaborated by his one-time student Francisco Varela. As he articulated a break with his former teacher, Hayles notes, Varela pushed autopoiesis to its radical conclusion by insisting on the distinction between a symbolic description and an operational one, a distinction that ensures that informatics remains conceptually distinct from the categories of matter and energy. Put simply, Varela believes that strictly speaking, neither information nor the "laws of nature" exist.[16] Ultimately, this led him to attempt to join autopoietic theory with the dynamics of self-organizing systems, with the aim of jogging the theory out of its "relentless repetitive circularity by envisioning a living organism as a fast, responsive, flexible, and self-organizing system capable of constantly reinventing itself."[17] In this respect, Varela's extension of Maturana's thought aligns with Hayles's own vision of a multiple and entangled relation between life and machines.[18]

The final phase of cybernetic research that Hayles recounts in *How We Became Posthuman*—virtuality—springs precisely from this entanglement and extends from the 1980s to at least the time of the book's publication. More precisely, this wave is the awareness that something *does* leap forth from complex systems (such as those made up of humans and machines) and that the exact nature of these emergent entities is constitutively unpredictable. For Hayles, this aleatoricism can be explained only through the development of a semiotics that is tailored specifically to the materiality of virtuality and therefore one that places the conventional paradigm of presence–absence in play with the informatic paradigm of pattern–randomness. Thus third-generation cybernetics furthers the discipline's movement away from the cyborg per se, directing itself instead toward the cognisphere.[19] In this light, Hayles strategically defines virtuality as "the cultural perception that material objects are interpenetrated by information patterns,"[20] a definition that reconnects perceptions of virtuality with the technologies that reinforce them. Through this connection, Hayles is able to unpack the seriated processes by which cybernetics has evolved (and continues to evolve), giving particular emphasis to the ways that material and conceptual motifs overlap, extend, and undermine one another.

Ultimately, then, the phases of cybernetic research around which Hayles constructs *How We Became Posthuman* are neither discrete nor exclusive of one another, nor do they preclude other, equally viable narratives. Instead, Hayles insists that, for example, first-generation autopoiesis remains active as a skeuomorph even after the discourse of cybernetics has shifted to the (virtual) terrain of artificial life and that this presence plays a larger role than simple nostalgia because it inflects the direction of future research.[21] Thus, echoing in advance the early-twenty-first-century "Macintosh revolution" in personal computing, Hayles is keenly attuned to the fact that our mode of interacting with a particular technology will shape its future development as much as the objective science itself does.

IN MANY RESPECTS, the critical encounters that Hayles incites in *How We Became Posthuman* do not come to full maturity until the publication of *My Mother Was a Computer* in 2005, where she examines the concept of emergence as it is narrated through what she calls the "Regime of Computation" (RoC). This worldview shares much with the reproductive logic of Darwinian evolution discussed in chapter 1 in that both emphasize the complexity that can arise from a "parsimonious set of elements and a relatively small set of logical operations."[22] In Hayles's view, this logic— which *is* computation—produces a Computational Universe, an ecology that resembles the realm of naturalized scientific privilege that Dyens performatively deconstructs. However, the two theorists confront this subject in different ways, employing different critical technologies: Dyens *extends* the logic of Darwinian (genetic) evolution, intensifying its underlying principles until they begin to disarticulate from their formational biology; in contrast, Hayles holds the complex dynamics of the RoC in tension, emphasizing the feedback loops that result from its working "simultaneously as means and metaphor."[23] Where Dyens emphasizes the process of Darwinian evolution to desublimate its tautologies and overdeterminations, then, Hayles's analysis of the RoC expounds the ways in which this logic sustains a dynamic equilibrium with existing constructions of materiality.

To begin her analysis, Hayles suggests that Steven Wolfram's Principle of Computational Equivalence best captures the RoC. In essence, this principle states that whenever behaviors in a system are not obviously simple, their sophistication corresponds to a computation of equal sophistication.[24] For Hayles, this principle contains three interlocking claims: first, that all complex behavior can be simulated computationally; second, that complex behaviors are computationally irreducible, in the sense that there is no way to render the labor of computing synchronically (e.g., in contrast to an equation); and third, that the process of computation does not merely simulate the behavior of complex systems but actually generates such behavior (in everything from biological organisms to human social systems).[25] The key point, for Hayles, is that the collective force of these claims lends the Principle of Computational Equivalence (and, by extension, the RoC) an ontological significance that it may not warrant.

This ground is claimed, Hayles argues, in part because of the metaphoric power of the RoC. That is, the RoC is culturally potent because its tropes tangle with social constructions of reality, resulting in "formulations that imaginatively invest computation with world-making power, even if it does not properly possess this power in itself."[26] The "in itself" of this formulation is crucial because it points to the importance of relationality to Hayles's thought: she does not argue that the force of computation is immaterial but rather that materiality has its own forces that inform computation (even as the latter constructs materiality), an understanding that leads her to remark that materiality is "the constructions of matter that matter for human meaning."[27] In particular, then, the RoC powerfully elides with our current computationally intensive culture to reinforce the latter's tendency to envision the universe as a giant computer, just as clockwork mechanism metaphors overdetermined the eighteenth-century European worldview.[28]

From this perspective, Hayles positions the RoC as a pathological metaphor—that is, as a symptom—and invokes Žižek's argument (from *Enjoy Your Symptom!*) to show that it is structured by the mechanism of teleological illusion, a mechanism that involves (in Žižek's formulation) "reasoning backward from one's present position [to see] prior contingent events as constituting a necessary and inevitable teleological progression to that point."[29] Thus, for example, Hayles argues that a feedback loop can be traced between the ever-increasing ubiquity and potency of computer technologies and the belief that physical reality is computational in nature. By contrast, she argues, someone like Wolfram fails to register this co-implication and instead assumes that the predicates of computation—its cultural, historical, and linguistic presuppositions—are objectively true. As a result, she notes, Wolfram's assumptions constitute a framework "within which problems are constructed and judgments are made"[30] without his even knowing.

In this view, the task of the would-be (cultural) analyst is to find a perspective from which the RoC can acknowledge that its truth claims are not objectively external to it. That is (continuing from the Žižekian formulation that Hayles invokes), the cause of a symptom is not found in reality per se but results instead from the parallactic nature of the Real, because "every attempt at symbolization is an attempt to suture an original cleft that is ultimately doomed to failure."[31] As such, "the subject's gaze is always-already inscribed into the perceived object itself";[32] one must identify with one's symptom, then, to recognize the bias in one's own subject position. Importantly, though, this distortion is also not objectively *internal* to the RoC: just as there is no "pure" object (i.e., no nonsubjective objectivity), there is also no subject without object. Thus, even when the RoC is apparently speaking only about itself (when it is speaking about specific computer technologies, for example, rather than about their ontological significance), there is an "objective, non-signifying 'reference'" relative to which it operates.[33] In this understanding, then, the RoC is paradoxically both internal and external to itself.

By thus positioning the RoC as a symptom of the perspectival limitations of an ambiguously constituted subject, Hayles offers an analysis that tries to account for the RoC as both means and metaphor.[34] Thus she argues that the RoC aligns with contemporary contestations of traditional metaphysics: in the RoC, everything from God to originary Logos to the axioms of Euclidean Geometry is allegedly supplanted. She notes, though, that this metaphoric critique is accomplished by *means* of a minimal ontology consisting only of an "elementary distinction between something and nothing (one and zero) and a small set of logical operations."[35] In this way, the RoC claims to avoid the ambiguity of deconstruction by closing the gap between an utterance and its effects: there is, seemingly, no *différance* between performative and constative claims in this understanding of computation.[36]

All told, Hayles is ambivalent about these claims. On one hand, it is clear that she is skeptical about the "new kind of science" that underwrites the RoC and that she is mindful of the politics that its metaphors perform. On the other hand, she writes from the perspective of someone situated in the midst of a culture that has inverted the erstwhile antiauthoritarian politics of postmodernism, an inversion that has yielded precisely the disembodied understanding of information that she critiqued in *How We Became Posthuman*. This depoliticization, she suggests, is due to the overwhelming privilege of attention afforded to signifiers over signifieds in deconstructive discourse, despite the fact that Derrida clearly argued against any such dominion.[37] Thus Hayles wages an affirmative action campaign of sorts, necessarily championing the importance of the signified to the worldview of code to correct the dematerializing tendency of signifier-laden contemporary humanities scholarship. All told, then, Hayles's entangled argument attempts to reaffirm a play of signs by emphasizing both the ways in which the material means of computation differ from other ontologies and the metaphoric implications of registering this materiality meaningfully.

Before proceeding in this light to Hayles's understanding of code, two important notes should be made about this method of argumentation (as she employs it): first, though Hayles is critical of the RoC, she mounts her critique against its ideology rather than its facticity; second, Hayles's position of critique leaves space for her to articulate the specific new problems posed by contemporary technologies and also to identify emergent horizons of possibility that are not found when either means or metaphors are considered exclusively.[38] From this, Hayles shows how opposed terms are implicated in one another as well as how this implication is productive of new understandings—particularly, understandings that can be attributed to the specific material conditions of computation.

This form of analysis (i.e., unpacking the co-implication of related but distinct elements) occurs frequently in Hayles's work but is only given an adequate name when she repurposes the term *intermediation* in *My Mother Was a Computer*. Specifically, intermediation, for Hayles, names the process through which multiple causalities, complex dynamics, and emergent possibilities arise from the interaction of media effects and a human life-world.[39] While intermediation in some ways resembles the notion of emergence—where discrete parts of an existing system combine unpredictably to produce a qualitatively different and more complex system—it differs from the latter by resisting the tendency to afford ontological priority to any of the actors in a system: whereas emergence tends to chart a unidirectional process characterized by an evolutionary trajectory, intermediation is figured through multidirectional and multicausal feedback loops that continually redefine their conditions of registration.[40] For Hayles, then, intermediation might thus be said to function as a narrative technology that facilitates her attempts to account for dynamic processes outside of the overdetermining logic of conventional causal language.

It is precisely this context that informs Hayles's analysis of code in *My Mother Was a Computer,* where she notes that the latter intermediates with the normative values of the systems that preceded it, namely, speech and writing. Thus, although she notes a "progression from speech to writing to code [wherein] each successor regime (re)interprets the system(s) that came before,"[41] Hayles insists that code does not "jettison the worldviews of speech and writing"[42] but instead remains in active interplay with them. However, Hayles additionally points out that in the same way that "writing exceeds speech and cannot simply be conceptualized as speech's written form,"[43] so, too, does code have its own characteristics that are not captured in either of these "legacy systems." Specifically, Hayles argues that code's excess to natural language is seen in the necessity of mediating between human and machine systems of meaning. That is, code is not—strictly speaking—intelligible to language since its complexity adheres in the "labor of computation that again and again calculates difference"[44] rather than in its original ones and zeros or in the operation of difference as such.[45] From this, Hayles insists that "code is performative in a much stronger sense than that attributed to language"[46] because it is *intrinsically* bound to materiality in a way that language is not.[47]

Put differently, the materiality of code means that it is site specific in its very syntax so that it is less vulnerable to the abstracting tendencies that frequently haunt language. This is the case, for Hayles, because code tends to be rendered unintelligible when it is transported to a different context.[48] That is, the context of a particular sequence of code is determined precisely by the level and nature of the code: the code and its context are often literally inseparable. As a result, she argues that code is not iterable in the Derridean sense in that it does not "carry with it a sign that breaks with its context."[49] Thus, although bits can be physically transported, their meaning cannot be, so that taking Derridean iterability to this level (which would correspond with the level of individual letters in language) would trivialize Derrida's argument more than it would say anything about code.[50]

Code's unique materiality is also observed by Hayles in the qualitative difference between the arbitrariness that it displays and that which Saussure attributed to the relation between the signifier and the signified. Here again, Hayles notes, "rules governing the transformations of encoding and decoding"[51] precisely specify the relation of the signifier and signified so that they act as material constraints that limit the range within which coded signs "can operate meaningfully and acquire significance."[52] However, this does not mean that a coded message is transmitted without deviation, for physical processes (and their concomitant aleatoricism) are still in play. Instead, Hayles's point is simply that these irregularities do not testify to the unavoidable undecidability that a deconstructionist reading would yield but rather to the inevitable presence of noise in the system.[53]

As such, there is a shift in the conceptualization of the sign that "signals a change of emphasis from the limitations of language in producing meaning to the limitations of code in transmitting messages accurately."[54] Thus the material difference between language and code speaks, for Hayles, to an ontological difference: "the arbitrariness of language implies the inability of language to ground itself in originary meaning, [whereas] the arbitrariness of code leads to multiple sites for intervention" in the process of coding, transmitting, and decoding information.[55]

Through this distinction (and others), Hayles concludes that the ontology of code ushers in a posthuman collective of encoders and decoders that displaces the unstable subject that existed with the age of writing, which itself shifted from the authority of a speaking subject.[56] Indeed, we might say that the collectiveness of this posthuman *is* intermediation in its (humanly) embodied form: paradoxically, by emphasizing that the machine is the final arbiter of computational intelligibility—"independent of what humans think of a piece of code"[57]—Hayles desublimates the interpenetration of language and code that is constitutive of each. By charting the commingled linguistic practices of humans and machines, she thus seeks after the specificities of each.

For example, Hayles interrogates the accretion of code and language in the relation between narrative and (relational) database.[58] Specifically, the dynamic character of relational databases—which, in the context of literal computation, have obsolesced their hierarchical predecessors—means that they operate asyntactically, in the sense that elements refer only to one another rather than to something external.[59] The flip side of this characterization, though, is that a database is limited to speaking only "that which can explicitly be spoken."[60] That is, not only does a database not refer beyond its own parameters but these parameters *are* the data themselves (rather than a predefined category into which the data are inserted). As is the case with code, then, the materiality of a given database is specific to its contents, which is to say that what a database signifies cannot be extricated from how it signifies.[61] In this sense, code and database are inextricable from one another and are formally interchangeable.

From this, Lev Manovich has (influentially) argued that narrative and database are "natural enemies" and has even expressed surprise that the former persists in the contemporary mediascape.[62] For Manovich, the database's ontological priority is evidenced by its ability to convert "sound, image, text, and their associated media [into its] digital code"[63] so that an object "consists of one or more interfaces to a database of multimedia material."[64] In this view, then, the database is paradigmatic but also material and real, whereas narrative is syntagmatic and dematerialized.[65] As such, Manovich views conventional linear narrative "as a particular case of hypernarrative,"[66] where the latter is not properly a type of narrative at all but rather an asignifying machinic process. Manovich, then, espouses a variant of the Computational Universe discussed earlier, positioning narrative as a subset of code wherein the latter is self-similar to the computational process that it performs (and that performs it).

By contrast, Hayles tacitly criticizes Manovich for falling into the structuralist trap of imposing a hierarchy based on formal operations and instead describes narrative and database as "symbionts" that, "like bird and water buffalo,"[67] together generate a complex ecology. Hayles's position manifests in the soft claim that databases require ordering to register outside of a machine, a point that also reiterates her argument for conceiving of information as always-already embodied. Additionally, though, understanding database and narrative as symbionts includes the strong claim that "narratives gesture towards the inexplicable, the unspeakable, the ineffable" and thus signal "more than can be indicated by a table of contents or a list of chapter contents."[68] That is, narratives always perform the inarticulable conditions of the causality that they claim, conditions that are always-already present in any causal relation. As such, narrative represents (for Hayles) the condition by which the database evades the overdetermining logic of simulation: precisely because narrative and database differ in their ontologies, purposes, and histories,[69] their complex intermediating ecology can emerge. Indeed, this perspective not only highlights the persistence of narration in contemporary database technologies but also emphasizes the database-like contingency of pre- and nondigital narratives.

THROUGH HER DESCRIPTIONS of the intermediating materialities of machine and flesh, Hayles's ultimate goal is to construct a version of posthumanism that is both intellectually and ethically tenable. As we have seen, this task requires her to reject the "nefarious" form of the posthuman that is "constructed as an informational pattern that [only] happens to be instantiated in a biological substrate,"[70] a vision that is shared by both Wolfram and Manovich (among many others), despite their very different worldviews. In place of this dematerializing logic, then, Hayles insists that computational processes always exist in relation to a human "legacy system."

And yet, if Hayles rejects technological determinism, she is also strongly motivated against liberal humanism (and, particularly, possessive individualism), which she sees as aggressively limiting the range of intelligible subject positions. That is, the liberal humanist subject views "human essence" as a form of self-possession that manifests as freedom from the wills of others.[71] As such, this subject operates in and as an economy of scarcity, privileging a form of possession that is situated—as Hayles and others have frequently noted—in and as a dialectic of presence and absence.[72] Ultimately, Hayles argues, this dialectic constructs boundaries that it then naturalizes—including subject–object, male–female, mind–body, and human–machine, to name only a few—resulting in the myriad exclusions and omissions that much of poststructuralism has sought to rectify. At the center of these binaries, though, Hayles sees a centered vision of the human as a unified force (i.e., human rather than nonhuman) that both preexists and persists beyond the distinctions that spring up around it: specifically, in the economy of scarcity that coincides with possessive individualism, materiality itself is restricted relative to the "full presence" of human agents.

As discussed earlier, though, contemporary information technologies do not "operate according to the same constraints that govern matter and energy."[73] As a result, Hayles is compelled to reject the human as an a priori and transcendental category: since it is no longer reasonable to sustain conventional conceptions of human agency, the (political) assumptions that underwrite "the human" are now visible whenever the category is called on. Thus Hayles's posthumanism is first a recognition of the historicity of the "human" so that her analyses all participate in a collective rethinking of it as an epiphenomenon. In this perspective, as in Derridean deconstruction, the liberal humanist subject is obsolesced because it is constitutively ahistorical: its very self-possession speaks to a part of that same self that evades possession and transcends history. In place of this subject, Hayles offers a posthuman that still engages the terms of subjectivity (including agency, consciousness, and will), but with a reflexive turn that avows the complexity of its dual status.

More accurately, this "dual status" is actually multiple because Hayles's posthuman subject is "an amalgam [that undergoes] continuous construction and reconstruction"[74] such that its autonomy is always undercut. As a result, Hayles emphasizes that (posthuman) cognition is distributed between "disparate parts that may be in only tenuous communication with one another."[75] If code and language are complexly entangled, for example, then thinking consciously in language implies a usually nonconscious thought that is externalized in machinic code. Importantly, this externalization is not simply metaphorical because thought itself is understood by Hayles as an embodied process so that our physical interactions with machines are literally conjoined with the machinations of our thoughts, which is to say our cognitive processes.[76]

In its most basic form, one can understand this observation intuitively. I might, for example, forget the numbers that are stored in my telephone's memory because I need only press the appropriate memory key to dial them. In this case, the telephone remembers the numbers for me; indeed, it is now a cultural truism that we forget a piece of information the moment that we enter it into a digital device that is designed to recall it for us at the appropriate time, be it a mobile phone, electronic address book, or digital calendar. In these scenarios, then, *memory* functions as a metaphor for the intermediating cognitive processes of humans and machines rather than referring to a recollection per se. That is, it isn't just that the device stores the given information—which, Bill Viola notes, has been accomplished by artificial memory devices for centuries[77]—it also reconceptualizes it in such a way that changes what counts as information in the first place.

That this is a shift in human *cognition* becomes clear when this tendency is contrasted with the long-observed pedagogical belief that writing something down enhances our ability to recall it: on one hand, written information is appended to a conscious thought as an enhancement device and is (in this sense) an autonomous and inanimate tool. By contrast, computationally distributed information is conjoined with conscious thought as a nonconscious memory that animates itself through its relations to other cognitive processes rather than predominantly through conscious commands. That is, as contemporary technologies combine and recombine to produce increasingly large and complex networks, it becomes less and less possible to convincingly promulgate a worldview that would posit a single human actor (or even human actors in general) at the center of these activities. Simply put, it is no longer possible to unplug from computer networks so that large networks (such as the Internet) have become bona fide ecologies in the fullest sense.[78]

Thus, though the example of digital organizational technologies is apt, it does not fully capture what is at stake in Hayles's approach; indeed, the fact that this example does not convincingly capture the distribution of causality that is implied by distributed cognition[79] speaks as strongly to our tendency to conceptualize causality relative to a human scale as it does to a deficiency in the example itself. This is the doubled ruse of digital consumer technologies: it is not only that these technologies are misrepresented as existing to fill certain human needs, whereas profit making is actually their raison d'être; although this is likely true, it is also the case that these technologies create needs in the first place. While this is the case for products of advertising in general—where advertising is typically understood to be an industry of need creation rather than fulfillment—the situation is amplified in the case of digital technologies because the generated need is a material one that exists in and as a causal nexus that cannot be reproduced metaphorically. The needs generated by new digitally synchronized technologies are not unreal in the same sense as, for example, those of a new miracle kitchen device. Instead, they are hyperreal in the sense that they are literally necessary to sustain the social fabric on which our subjectivity depends. Thus, for example, France's constitutional council recently struck down a "new law which would have allowed the state to cut off the Internet connections of illegal file-sharers for up to a year," noting that "'free access' to online communications services is a human right."[80] Stated in the language that we will encounter with Mark Hansen in chapter 5, the point (which is also Hayles's point) is that the body's coupling to its external environment, while always potentially technical, "can increasingly be actualized only with the *explicit* aid of technics,"[81] and this explicitness constitutes an important material change. In short, the distributed cognition of contemporary culture means that we live in an ontology of distributed causality so that to recapture digital technologies in a language of linear causality (of which advertising is an example) is a fundamentally conservative ideological gesture. Simply put, the needs—and the accompanying obsolescences—generated by digital technologies are material rather than primarily materialistic.

Hayles's insistence on this multicausality lies at the heart of her dispute with psychoanalysis. Specifically, Hayles criticizes Žižek for investing everything in "a theory of the unconscious based on [our conventional] physical and mental structures"[82] and thus failing to recognize that human action is "coordinated with complex virtual/actual environments characterized by flows and relations between many different agents—including non-human ones—tied together through distributed cognitive networks."[83] This is presumptuous of Žižek, Hayles insists, because it "holds the entire span of the far future hostage"[84] to present local conditions, which include the (embodied) cognitive processes that accompany our movements through the world. Simply put, Hayles finds Žižek's psychoanalytic approach to be dependent on an ahistoricized vision of the human that is constrained by analog (and anthropocentric) constructions of causality.[85]

The most significant such construal of causality that Hayles finds in Žižek (and Lacanian psychoanalysis in general) is the dialectic of presence and absence. For Hayles, understanding subjectivity through this lens dooms psychoanalysis to recapitulating the predicates of possessive individualism, even though it is clearly motivated against the latter's claims (i.e., claims to self-possession and, by extension, to full presence). That is, Hayles's perspective demonstrates that while Žižek succeeds in articulating the radical contingency of (human) agency, he nonetheless situates this contingency within the material and temporal scales of analog human cognition. Citing Guattari, Hayles suggests that this neglects the heterogeneity of machines, which refuse to be "orchestrated by universal temporalizations."[86] In so doing, Žižek is seen (by Hayles) to be locked into a particular way of thinking that profoundly disavows nonhuman actors as well as relativizing what is radically nonhuman in "human" actors. The "human" that is thus privileged takes on the status of an alibi, then, appearing to be neutral while actually acting as a metaphor for a construction of the human—a territorialized assemblage—that reinscribes the formational logic of the possessive individual.

For example, consider Žižek's understanding of Lacanian "forced

choice," which "consists of the fact that the subject must freely choose the
community to which he already belongs, independent of his choice."[87] We
can understand Hayles's departure from this perspective by considering the
specific situation of the paradox in this formulation. For Žižek, the paradox
comes about through the impossibility of epistemologically encountering the ontological Real, where the latter "has no positive-substantial
consistency [and] is just the gap between the multitude of perspectives
on it."[88] By contrast, Hayles's worldview sees the community (or, more
accurately, the communing) and the subject's choosing acting together so
that the paradox speaks to the complexity of their interactions. In Hayles's
perspective, then, Žižek neglects the subject's implication in communities
that bear no relation to her choosing, which is to say that he ignores the
fact that communities consist of intermediating materialities, temporali-
ties, embodiments, and so on. Whereas Žižek characterizes the Real as
a paradoxical series of incommensurable relations, then, Hayles views it
as a commingling of qualitatively different elements.

In this context, pattern–randomness operates in Hayles's thought as
a means of thinking distributed cognition, an achievement that would
not be possible within the collection of overdeterminingly analog metaphors that drives the presence–absence dialectic. Following Guattari,
Hayles argues that "semiotics has falsified the workings of language by
interpreting it through structuralist oppositions that covertly smuggle in
anthropomorphic thinking characteristic of [a conscious human] mind."[89]
Since our inscription as actors in distributed cognitive environments renders us hybrid entities "whose distinctive properties emerge through our
interactions with other cognizers within the environment,"[90] introducing
pattern–randomness as a term in the discourse begins to account for the
currently underrepresented specificities of information, specifically, its
emphasis of access over possession.[91] By invoking machinic operations
(through pattern–randomness), then, Hayles makes available "a materialistic level of signification in which representation is intertwined with
material processes."[92] Here again, the important element is the complex
interactions that emerge from and within these dialectics rather than the
(seemingly) discrete actors that partake in them.

In particular, what emerges from Hayles's account of cognitive distribution is an evolved notion of machine agency, which is not rendered

in the sci-fi terms of machines gaining consciousness and overtaking the world but rather in the sense of machines making choices, expressing intentions, and performing actions.[93] In this, Hayles argues that "if the posthuman implies distributed cognition, then it must imply distributed agency as well, for multiplying the sites at which cognizing can take place also multiplies the entities who can count as agents."[94] In the case of machines, these agencies are often expressed in the scale and syntax of code so that they are translated (in the full sense of the term) when perceived by humans.[95] As such, machines are both agential in their own right *and* in the assemblages that they form with human subjects, and these two intensities are discrete symbionts. In the fullest sense possible, then, Hayles's technological posthumanism "does not require the subject to be a literal cyborg" for it to be deeply interpellated by digital technologies.[96] Furthermore, even literal contact with the technologies themselves is not entirely necessary to register a cybernetic shift in human subjectivity since human actors are continually subject to the secondary effects of machine actions. In this view, agency—long identified with free will and a rational mind—becomes "partial in its efficacy, distributed in its location, mechanistic in its origin, and bound up at least as much with code as with natural language."[97]

Indeed, Hayles notes that cybernetic discourse had a profound historical influence on (and was influenced by) numerous key poststructuralist thinkers, including Lacan, Deleuze, and Guattari, who all "use automata to challenge human agency."[98] Moreover, these thinkers (each in his own way) articulate what Hayles calls a "crisis of agency," showing that the challenges posed by machines to humans flow both ways, configuring automata as agents. Summarizing this position, Hayles notes in *My Mother Was a Computer* that "if, on the one hand, humans are like machines, . . . then agency cannot be securely located in the conscious mind. If, on the other hand, machines are like biological organisms, then they must possess the effects of agency even though they are not conscious."[99] In particular, she notes that Deleuze and Guattari envision the unconscious as a program rather than as a dark mirror, a shift that "can scarcely be overstated, for it locates the hidden springs of action in the brute machinic operations of code":[100] mechanistic, computational, and nonconscious operations that "nevertheless display complex patterns that appear to evolve, grow, invade new territories, or decay and die out."[101] In light of these claims, then, we might presume that the contemporary technological posthuman emerges not only from the interaction of machines and humans but also from the intermediations of their respective discourses.

Hayles's technological posthuman is "post" not because it is unfree from the "wills of others" but because there is "no *a priori* way to identify a self-will that can be clearly distinguished from an other-will."[102] That is, Hayles's reading of technological posthumanism seeks to counterbalance the liberal humanist subject that it is mobilized against, transforming the latter's locatable cognition and untrammeled free will "into a recognition that agency [and cognition are] always relational and distributed."[103] Clearly this shift is motivated by Hayles's ethical and political concerns and constitutes an attempt to situate technological posthumanism outside of the computationally dematerializing language of technological determinism while also registering the material specificities of contemporary technologies. To simply oppose the two ontologies—human and machine—would only be to reveal "a ghost facing a skeleton."[104] Instead, Hayles both aims to uncover the "software ideology"[105] of the RoC and to be cognizant of the nefarious market relations that permeate the discourse of possessive individualism;[106] however, she is most interested in the specific processes through which they interact. For Hayles, then, an essential component of coming to terms with the ethical and political implications of contemporary technologies is recognizing the codependence of humans and machines as well as the extent to which each is involved in making the other.[107]

However, although articulating an ethical version of technological posthumanism is purportedly one of Hayles's fundamental projects (and informs every aspect of her writing), what precisely this ethics consists in remains unclear. That is, although the reader can infer a certain set of ethical beliefs from Hayles's writing, she is left to deduce on her own how these notions combine to give an account of ethics itself. This absence is notable because it is difficult to imagine how ethics might be articulated without prioritizing the human in a way that Hayles's posthumanism otherwise resists.

Of course, there do exist nuanced notions of ethics that might reasonably apply to Hayles's worldview, many of which involve a responsibility to the Other. Derrida, for example, stages ethics (a term he uses reluctantly) as "a product of deferring, and of being forever open to possibilities rather than taking a definitive position."[108] However, I would suggest that Hayles's insistence on the positive dimension of intermediation (i.e., on the joining of means to metaphor) at the very least troubles the reader's ability to assume that this is the kind of ethics she has in mind: to the extent that machines and machinic processes are attributed agency, what constitutes the Otherness to which they bear responsibility? That is, what is radically heterogeneous to the worldview of intermediation?

In fact, Hayles does not address these questions, and her decision not to do so points to another sense in which the desire to articulate an ethical version of technological posthumanism drives her thought: rather than pushing through on the strength of her (machinic) convictions, she instead withdraws to an assumed (and therefore uninterrogated) ethical ground. Specifically, this ground consists in many of the assumptions to which Hayles otherwise objects: while she distances herself from the liberal subject (which, as mentioned earlier, often includes the Lacanian subject in her reading), her thought continues to be haunted by a notion of humanity that is heavily influenced by the precepts of liberalism. That is, isn't it the case that an untroubled notion of human individuality continues to provide the basis from which Hayles evaluates posthumanism, even while she stages the latter as a deconstruction of human centrism?

In fact, this appraisal actually mirrors a form of critique that Hayles frequently directs at others in her writing. For example, when calling dominant approaches to literary criticism print centric, she notes that print provides the baseline for critique, even though the critiques are postulated as including other media as well.[109] Thus, though Hayles is alive to the incommensurable materialities of humans and machines (as well as to the permeability of the border that separates the two), this does not necessarily translate into a nonanthropocentric ethical project: simply broadening the scope of what might be considered agential does not necessarily move Hayles beyond the pale of human centrism (nor does it describe what such a "beyond" would entail). In fact, the reverse may even be the case in that Wolfe has noted that one of the hallmarks of humanism (and especially liberalism) is its "penchant for that kind of pluralism in which the sphere of attention and consideration (intellectual or ethical) is broadened and extended to previously marginalized groups [without destabilizing] the schema of the human who undertakes such pluralization."[110]

As a further—and more concrete—example of how humanism continues to inform Hayles's ethical worldview, consider the ambiguity of the term *meaning* in her writing. Curious about this, after a public lecture I asked her what nominates it as a privileged term and what that privilege might foreclose.[111] In a thoughtful and detailed response, she agreed that "meaning obviously occupies a privileged place in [her] thought" and even suggested that humans might be defined as meaning-seeking—and often meaning-constructing—animals. However, she continued by saying that there are distinct differences between the various versions of meaning and that she "wants to break the hegemony of consciousness's grip on meaning, [which exists as though it] were the only site at which meaning-making can take place."[112] Thus it is important to Hayles that the term *meaning* does not specify what the context is in which meaning can be acquired, which is to say that (for her) meaning can register independent of humans. Nonetheless, that meaning making does in fact occur is one of Hayles's "fundamental beliefs" and is part of her view of who she is and of what humanity is.[113] In short, then, Hayles effaces meaning as a linguistic actor but nonetheless relies on it in her texts' performance.[114]

If meaning making is privileged, even in the indeterminate terms that Hayles offers, the question obtains, what does this privileging work in service of? In most cases, the answer to this question points back to the argument that Hayles remains more ensnared in the language of humanism than she allows. For example, consider again her claim that narrative and code are symbionts, involved in a complex relation (in the full sense) where each is of equal importance: if this were literally true, then encountering code *in its own terms* would be to render it immaterial—given her argument that binds materiality to *human* meaning[115]—and this immateriality would fold back into the human in the same way that automata are configured as agents in an environment of distributed cognition. That is, contrary to Hayles's stated objectives, this view constructs code as a force of dematerialization. The point, then, is that privileging meaning means that the term means something when it is privileged, in this case signifying a hegemony that flows from the human.

In contrast, if we assume that the human has an (unspoken) ethical

priority in Hayles's thought, her claims resonate much more sympatheti-
cally with one another. For example, she suggests that "our narratives
about . . . virtual creatures can be considered devices that suture together
the analog subjects we still are, as we move in the three-dimensional spaces
in which our biological ancestors evolved, with the digital subjects we are
becoming as we interact with virtual environments and digital technologies."[116] In the notion that narrative acts as a suture, the human itself is
given ontological precedence: remembering that narrative is a human
mode of signification for Hayles, the general conditions of human meaning (narrative) are necessitated by the gap between particular narratives
and the databases on which they draw. In short, the "intermediation" of
narrative and database is grounded in a human narrative, which grounding constructs the perspective from which the intermediants are viewed.
In a sense, Hayles's own critique of Manovich—her argument against his
hierarchical structuring of narrative as a subset of database—turns back
on her in its inverse form; thus flipped, the logic of the critique still obtains
because it reveals that Hayles has imposed a structural hierarchy where a
symbiotic relation was promised.

This is further exemplified by Hayles's proposal in *My Mother Was
a Computer* to "regard the transformation of a print document into an
electronic text as a form of translation," a task that includes rigorously
specifying what is gained and lost, what these gains and losses entail, and
"especially what they reveal about presuppositions underlying reading and
writing."[117] Here again, the questions arise: specify from what perspective?
Are there simply "changes" to the text, or does the notion of change itself
need to be brought into play? The simple fact is that Hayles's resistance to
full-blown constructivism necessarily results in her having to choose a scale
and perspective in which the "physical characteristics" that interact with
signifying structures to produce materiality[118] can be registered as such.
Since this "choice" is constitutively irrational, inexplicable, and beyond
the grasp of her subjectivity (because it grounds her subject position and
because it is not supported by any original claim), it ultimately consists
in a claim to authority that is masked by the fact that it "makes sense."[119]

Similarly, Hayles's discussion of morphological resemblance (i.e.,
similarity of form) also gains traction through the prerogative that she
grants to anthropocentric meaning. Specifically, she argues that the analog

function of morphological resemblance is "the principal and indeed (so far as [she knows]) the only way to convey information from one instantiated entity to a differently instantiated entity."[120] Here again, a preexistent notion of meaning frames the argument, indicating what constitutes an "instantiated entity." If, for example, we took her constructivist definition of materiality without its hegemonized notion of meaning, then this claim about morphological resemblance could be debunked simply through reference to the imagination (or hallucination): one imagined entity can communicate with another without referring beyond the closed circuit of communication. Instead, morphology enters into the picture precisely because meaning, for Hayles, includes an unacknowledged privilege that is afforded to the (normative) reality-scale of individual humans. Thus, when she suggests that whenever "narratives cannot be constructed, the result is . . . a world of inexplicable occurrences,"[121] we can note that explication—an unfolding of meaning—presumes its own necessity in her thought. Simply put, meaning is an instance of teleological illusion and is thus (by her own reasoning, as discussed earlier) symptomatic of a deeper conflict. That is, remembering that *différance* is always operative in and as analog reality, the fact that analog resemblance is the "dynamic that mediates between the noise of embodiment and the clarity of form"[122] signifies an injunction to mean that, at the human level, operates as the condition of registering meaning in general. The argument that materiality is "the constructions of matter that matter for human meaning" thus flows both ways; if analog reality is both the form of mediation and that which is mediated, then that which is inexplicable to it can only be registered through a process of violent reduction (i.e., the act of registration itself).

To be clear, while an abstract human individual manifests in the metrics that obtain in Hayles's work, it is to her great credit that this is much less the case at a local level. That is, I do not wish to discount the extent to which Hayles does succeed in repositioning meaning in her attempt to account for the feedback between humans and machines. As she insists, meaning remains operative, but it is appropriated for a "much broader view of cognitive and sub-cognitive processes."[123] What this view entails, importantly, is the possibility of claiming that cognition is computational—in the sense that it generates a subatomic level—"even while conceding differences in embodiment and the integral relation between embodiment and human cognition."[124] Hayles's point, then, is ultimately a political one, couched as it is in a language of analysis; scilicet, her work might be summarized as performing the dual insistence that the ubiquity of computational processes is a *fait accompli* manifested at every conceivable level of reality but also that this computational omnipresence need not produce the particular political configuration assumed by the RoC.

ULTIMATELY, HAYLES'S THOUGHT is whipsawed by the challenge of radical thought that Baudrillard posed, the challenge to make the world, which is given to us in unintelligibility, "more unintelligible, more enigmatic, more fabulous."[125] On one hand, this is clearly where her ambitions lie: her insistence on the indeterminacy of intermediation, the incommensurability of code and language, and the impossibility of abstractly registering embodied processes (i.e., turning them into data) speaks to this comportment. And yet radical thought ultimately remains beyond her because she will not (and perhaps should not, for that matter) cede a presumed ethics that is formational to her project. As such, her thought remains within the human values that she seeks to replace: there is no denying, for example, the principles of computational reality and structures of computational exchange that she claims, but she is unable to accept that computationality itself—taken overall—cannot be exchanged for anything human. Hayles recognizes this unintelligibility of code in its specific, embodied exchanges with language, but she nonetheless exchanges code itself for human (narrative) value. That is, the radicality of code—its collapsed space and time, its immediacy, the fact that it constitutively does not lack or desire—disappears into Hayles's (humanist) ethics, where it becomes an object of a choosing subject. In short, though Hayles understands embodiment as a radically dynamic process that is inclusive of—and also, in a sense, driven by—technologically distributed cognition, the subject (perhaps posthuman but also importantly human) nonetheless continues to act as the ultimate ethical arbiter in her schema.

This, in itself, is not a bad thing, and it is difficult (and perhaps undesirable) to argue against a politics that seeks to give voice to those who have long been denied one. For example, autistic people, in Hayles's view, are given an intellectual cachet that could not be registered without understanding the interplay of code and narrative.[126] Clearly this is a real and consequential shift in humanist discourse, one that extends the values of inclusivity that most readers—and I—generally hold in high esteem.

Moreover, Hayles's technological posthumanism tempers the psychic and physical pains that can accompany human reliance on computation by bringing the anfractuosities of narrative to bear on problems that were previously thought computationally. For example, she discusses the experiences of a software engineer designing a system to help deliver information to AIDS patients: by meeting the patients for whom the software is designed, the engineer realizes the extent to which the representation of information in the program reduces the material concerns of the patients. Thus, for example, the programmer recognizes that what is for her a data set of tuberculosis symptoms represents for her clients the "complicated biological urgency" entailed in the "fear of sitting in a small, poorly ventilated room" with someone who has a medication-resistant disease. In turn, this recognition allows her to redesign the software with this in mind.[127] Ultimately, then, Hayles's analysis wagers that an adequate representation of the problem (i.e., understanding it as symptomatic of the contrast between the worldviews of code and human language) will allow other people working at the intersection of humans and machines to learn from the software designer's experience.

That said, though, Hayles is unable to represent this same problem within the ethical scheme of her construction of technological posthumanism so that, couched as it is in a language of analysis, this ethics is naturalized. As a result, Hayles's "goodness"—the general admirability of her project—is situated in and as her blind spot, simultaneously validating and limiting the arguments that she mounts. If it is clear that a particular understanding of meaning is at the center of Hayles's notion of the human, then, it is also clear that via this understanding the human is installed at the center of her construction of technological posthumanism. That is, at the very least, the human serves as a baseline for Hayles's definition of the posthuman so that, for example, she is not moving toward the genetic stratum as the basis for an ethical comportment (as Dawkins does).

In this context, consider that the crucial question of *My Mother Was a Computer* is "how the 'new kind of science' that underwrites the Regime of Computation can serve to deepen our understanding of what it means to be in the world rather than apart from it, comaker rather than dominator";[128] in this formulation, who are "we" if not humanity, in the most conventional sense? No longer naturalized under the sign of the human, Hayles's technological posthumanism marks the point where the human loses its originary status and becomes a subject of discourse, but also—paradoxically—the point where human meaning authorizes itself by forming the ground of this same discourse. In this sense, then, Hayles reinscribes the human in her technology of the posthuman: she forecloses the "radical material exteriority" of technology in favor of a "merely relative exteriority."[129]

IN FACT, the uninterrogated political bent of Hayles's own thought is conveyed in its means, in the way that it gains traction. After all, what is at stake in the shift in Hayles's attitude toward post–World War II continental theory that takes place between *How We Became Posthuman* and *My Mother Was a Computer*? While her early, often scathing, analyses of Foucault, Lacan, and Derrida are usually (if not always) convincing, what does Hayles gain by citing these thinkers, specifically, in a problematic conflation of possessive individualism, liberalism, and Lacanian psychoanalysis? Moreover, is there something gained that she does not then give up when, in *My Mother Was a Computer*, she retracts much of her criticism and once again disarticulates these thinkers from one another? If the human naturalizes a certain conservatism in Hayles's ethical and political assumptions, might this shift in her thought act as a twin movement that overemphasizes the radicality of her position?

Truthfully, there is no way of knowing the answers to these questions because they point to the machinations—the means—of discourse itself, which displays its own logic of distributed causality. And yet it does bear noting that Hayles gains traction in *How We Became Posthuman* based on the appearance of a radical politics that would resist the (masculinist) overdetermination of technological discourse not just from a feminist perspective but also from the perspective of machines themselves. In her account of Janet Freed–Freud, for example, Hayles emphasizes the coupling that takes place between the conference assistant and the transcription technologies that she uses, noting that even though Freed has no voice in the transcripts, "the transcripts have a voice that we can only read because of her."[130] From this, Hayles draws a connection between the erasure of Freed's embodied activities (as an amanuensis) and the dematerialized vision of materiality that the conferences helped to disseminate. The mistake, in Hayles's view, is the abstract, idealized understanding of technology that this suggests. Thus Hayles's analysis includes the corollary claim that understanding the material processes of the technologies in play in a given situation—in this case understanding that information technologies, typewriters, and human actors cooperate—offers an opportunity for a politics of equality and inclusivity. This perspective, in turn, is supposed to be radical in that it is staged as a departure from the disembodiment in which Hayles believes poststructuralist thought (in the broadest sense) participates, despite its claims to the contrary.

This departure is resituated, though, when Hayles emphasizes the role of many of these same thinkers in the development of cybernetic discourse. If the move in *How We Became Posthuman* is to inject the materiality of machines into the (purportedly humanist) discourses of subjectivity emblematized by Deleuze, Guattari, and Lacan, then *My Mother Was a Computer*'s resituating of all three as thinkers of a posthuman subject "in which consciousness, far from being the seat of agency, is left to speculate why she acts as she does" would seem to alter retroactively the significance of the initial thrust.[131] My own intention here is neither to agree nor to disagree with either of these analyses; instead, I only want to note that Hayles retains the claim to radicality of her initial encounter with these thinkers, but this radical aspect is detached from the content that originally defined it. As a result, what remains is by definition a simulated politics of change that perhaps reinscribes the values against which it is aimed, this time with the added authority of having killed its theoretical father. In a different sense than was suggested with respect to "cultural bodies" in chapter 1, then, the humanist technological posthuman might be considered hyperhuman: not born through the detachment of genetic processes from their biology but rather as the turning back on itself of human meaning. The reversal potential of Hayles's recalcitrant insistence on the specificity of embodiment, then, may be the disembodied abstraction that haunts human meaning as its necessary formal component.

IN HER MOVEMENTS away from both possessive individualism and techno-logical determinism, Hayles succeeds in revealing their implication in one another. Both perspectives dematerialize, decontextualize, and disembody while simultaneously naturalizing visions of materiality, and Hayles's oeuvre provides a long overdue corrective to these tendencies. Moreover, her mobilization of intermediation—replete with a robust theoretical lexicon and wide-ranging disciplinary influences—offers an important response to the challenge of thinking subjectivity in the contemporary media ecology. Revealing the presuppositions of technophilia and technophobia alike, Hayles's humanist technological posthumanism is tuned to the privileges, exclusions, and prejudices that haunt contemporary technologies and shape the everyday activities of a flesh-world that is increasingly difficult to extract from them.

As such, humanist technological posthumanism is fundamentally a revision of human identity that presumes that such revisions are possible and, indeed, inevitable. Moreover, these texts are infused with the de facto belief that there is a better and worse way to proceed in this respect and that these events are subject to ethical, political, material, and intellectual will. Hayles repeatedly insists that "what we make and what (we think) we are coevolve," noting that "the parenthesis in the aphorism marks a crucial ambiguity, a doubleness indicating that changes in cultural attitudes, in the physical and technological makeup of humans and machines, and in the material conditions of existence develop in tandem."[132] Clearly Hayles aims to effect such a change.

This pursuit is of no small consequence. In what is often regarded as a culture of completed capitalism, material specificity of the type that Hayles advocates reverses the flow of universal exchange, the reduction of bodies to data, and the sanitization of events rendered in the language of probability. Against these tendencies as they are found in certain readings of (first-generation) systems theory, then, Hayles offers narrative as a "more embodied form of discourse."[133] However, the fact remains: since values are always symbionts with their material instantiations and since Hayles's values coincide with those of specifically human meaning, then we can only presume that the metric of embodiment in Hayles's writing boils down to an abstract human body, a body that Hayles herself has shown to reduce normatively the definitions of who can speak and what can be said. Whether this can also be the site of a progressive politics, as Hayles hopes, remains to be seen. If this is the case, though, it will be accomplished through a posthuman politics that is first and foremost a restaging of humanist values. As such, the resulting human recombinant will necessarily be faced with the challenges that have dominated discourses of subjectivity ranging from liberal humanism to Lacanian psychoanalysis to cybernetics, so that its political gains will also necessarily place restrictions on who—and what—is granted the privilege of pursuing such lofty goals.[134]

4   *THE TRACE*: MELANCHOLY AND POSTHUMAN ETHICS

> The production of the *un*symbolizable, the unspeakable, the
> illegible is also always a strategy of social abjection.
>
> JUDITH BUTLER, *BODIES THAT MATTER*

The year 1995: it was a time of flesh, it was a time of data. It was a virtual
time, really, in the full sense of the term, that is, in the sense that data bodies
were seen circulating freely, but their visibility as such attested to an explicit
awareness that something was not circulating. Isn't this the story of virtual
reality? That every dematerializing claim that it makes is simultaneously a
demonstration of the necessity of making such a claim, a demonstration
that precisely the opposite is also true? We encounter Rafael Lozano-Hemmer's *The Trace,* then, as volume 1 of "The History of *Virtuality,* echoing
Foucault in a time when the unflagging claims of the flesh's "disappearance" into digitality spoke equally to a relentless normative reiteration of a
historically specific version of the human body. *The Trace,* then, is read here
neither as the author of a dematerialized body nor as evidence of embodiment's continued relevance (à la Hayles). Instead, *The Trace* unilaterally
reduces its participants' modes of relating to one another, but this reduction in turn performs a critique of both vectors of (de)materialization and,
ultimately, positions their relation as the terrain of its posthuman ethics.

The Trace: Remote Insinuated Presence consists of a telepresent installation that invites two participants in remote sites to share the same
telematic space.[1] The piece features two stations (which can be in the same
building or in different cities), each of which consists of a dark room with
a giant rear-projection screen on the ceiling, a side monitor, four robot
lamps hanging from the ceiling, and ten loudspeakers distributed around
the room's perimeter. On entering the station, each participant is given a
small wireless sensor that monitors her three-dimensional position. *The
Trace* transfers the sensor's coordinates between the remote stations so
that each sensor controls audiovisual elements in both stations.

The Trace insinuates the remote participant's presence in the local station in four ways, first, via four mechanized robot lamps that hang from the ceiling, each emitting a narrow beam of light (two beams are yellow and two are blue). The blue beams point to the local participant, whereas the yellow beams intersect at the position of the remote participant (fog machines are used to accentuate the beams). Second, three-dimensional graphic animations of a ring and disc are projected onto a large ceiling screen, with the ring tracking the movements of the local participant and the disc following the remote participant's movements. Third, the remote participant is rendered locally through quantitative data presented as a statistics screen on a CRT monitor. Finally, a perimeter of loudspeakers articulates sounds that are spatialized relative to the position of the remote participant. In addition, the sound volume is adjusted in direct relation to the distance separating the participants, whose proximity to one another is also reflected in the content of the sound itself. Specifically, the distance is quantified in meters, with a prerecorded voice speaking the distance triggered appropriately (e.g., when the two participants are one and a half meters apart, a voice speaks "one point five meters").

In the context of mid-1990s digital art, the multisensory aspect of this work is notable because it offers a multiply embodied telepresence. In this, the work resists the visual domination that, for example, characterized early video games, which were typically configured around the represented movements of an avatar (i.e., nonvisual media were typically put in service of the visual narrative rather than considered in their own right). Instead, *The Trace* works not only in multiple media (i.e., sonic, visual, and locative)[2] but also in multiple registers within each medium. Thus, for example, participants have simultaneous locative experiences in the two-dimensional world of statistics, the three-dimensional world of the robot lamps, and the simulated three-dimensional world of the projected graphics. Through the interactions within each medium, then, an internal complexity is produced that, in turn, renders the participants' movements with a degree of specificity—a physics of sorts—that was not available in contemporaneous computer games.[3]

However, in comparison to contemporary telepresent technologies, *The Trace* could nonetheless be read as perpetuating precisely the dematerializing narratives that N. Katherine Hayles critiques so thoroughly.

FIGURE 4. Rafael Lozano-Hemmer and Will Bauer, *The Trace, Remote Insinuated Presence* (Collection Fundación Telefónica, Madrid, Spain), 1995. Photograph by Antimodular.

Certainly the piece is not realistically immersive to the extent that today's technologies allow, which is to say that the parameters for translating the participants' actions into quantifiable data do not approach the level of sophistication required to make the translative process "disappear." Put differently, the participants are computed by the piece rather than existing in a fluid exchange with it, such that—from a basic technical perspective— the aesthetic interest of the environment consists primarily in clarifying the visitor–system correlation. In short, *The Trace*'s immersion does not possess the continuity that is necessary for a digital environment to be "believable" so that its status—in this understanding—remains objectively figured (i.e., rather than existing as an interface).[4]

This line of critique seems disingenuous, though, not least because it suggests the type of unilateral technophilia against which *The Trace* is motivated, particularly since Lozano-Hemmer is clear that the multisensory conditions of the piece are not intended to work toward a realistic representation of embodied reality. Instead, the crucial aspect of the piece's multisensoriality is that it is multiplicitous: the logic of the various representational media yields qualitatively different representations that each act within separate—but interacting—virtual architectures. Thus, for example, while the graphic projections and locational sound each correspond to the participants' movements, there is little relation between the actual representations themselves: the graphic moves over a two-dimensional space (the ceiling) that is at a distance to the participants, and its depth is articulated through simulated shadows, whereas (by contrast) the sound immerses the participants by filling the whole room and registers their location through sites of intensity. Thus, whereas the graphic absents a given position when it moves to another, the sound is ubiquitously present as a composite of relative intensities.[5] This difference not only suggests a qualitative difference between the two media but also highlights the use of different interpolation ratios—and thus the creation of differently "sized" virtual spaces—both between media and between spaces.[6] As a result of these differences, the interpolative process is foregrounded over its results.

This has two distinct effects: first, because a single data source (i.e., each participant's tracking location) is translated across multiple media, the medial fungibility of digital data is emphasized. Second, because these media are continually returned to the participants as their own embodied experience—that is, because the media are the participants' embodiment in the context of the piece—the gaps over which data's "a-mediality" sutures are made visceral in a way that could not occur if only a single medium were in play.[7] The multisensory aspect of the piece, then, introduces a reflexive quality to the telepresence experience, wherein the representational media point the participants' senses back to themselves: they foreground the "complementarity between informational environment and disembodied embodiment."[8] *The Trace*'s telepresence experience is not so much one of extending the body within conventional parameters, then, as it is one of reconsidering the ontological presuppositions that underwrite these conventions in the first place.

Nowhere is this more evident than in *The Trace*'s sound, which literally gives voice to the mediating agency of the piece. That is, the (tele)presence of a speaking voice suggests a machinic agency because it articulates the mediated distance (or *teledistance*) between the two participants from a perspective that can only be that of the computer: the spoken distances—"one-point-five meters," "one meter," and so on—only exist in the computer's topology. In this way, the machinic agency of the voice sutures the opposition that is found in the immediate content of the sound: on one hand, environmental sounds are ever present and register the remote participant's movements through continuous dynamic panning; on the other hand, the text is spoken in discrete intervals and offers measurable content. By lending a material presence to the machine, then, the voice (rather than the text it speaks) suggests that this opposition is itself grounded in a relational mechanism. Simply put, the opposition (between discrete and continuous sound) exists as an opposition only within the context of a preexisting field of mediation.

In this way, the disjunctions between and within media highlight the role of representation in the piece, which functions to disarticulate each participant's body from itself. Thus, if the entangled multimedia rendering of the participants reveals the reduction–addition that Hayles finds in all media translation, it also—more importantly—suggests that the participants' bodies are no longer unified sources of technological extension but are rather implicated in complex medial relations that both exceed and interpellate them. Simply put, the participants are not able to control their actions—or, perhaps more truthfully, they are no longer able to maintain the illusion of control—because the technology in *The Trace* renders each seemingly discrete action as a multiplicity of incompatible actions.[9] The participants' disjunctive digital embodiment does not point to their flesh bodies as the "final instance" of their selves, then, but instead registers the multiplicity of their (nonetheless constrained) agency.

The centrality of agency—as a problematic—to *The Trace* is evidenced in Lozano-Hemmer's characterization of it as an experiment in "virtual lebensraum," a test of whether "the physical distance we are expected to keep from other people is upheld within telematic systems."[10] The artist's objective in the work is to allow the participants the possibility of tele-embodying, which is to say, the possibility of occupying "identical positions in telematic space to the point where they are inside each other."[11] This goal is signaled by the effects that are triggered when tele-embodiment occurs, which include dramatic movements of the robot lamps, an animation in the projected graphics, and flooding of the space with sound (where "flooding" is executed by increasing the intensity of the sound while simultaneously decreasing the differences in intensity between the individual sound channels).

In a sense, then, *The Trace* folds data back into the regime of desire that has often dominated subjective discourse and questions of agency. That is, the individual participants' agency that was disinvested by the piece's multisensoriality is reinvested in their (desiring) telematic relation to one another.[12] *The Trace* does not so much dematerialize the participants' bodies nor extend them beyond the constraints of presence and absence; instead, the piece operates within a paradigm of social agency in which the participants' relations are foregrounded over their embodiments. Simply put, the primary aesthetic territory of the piece is not embodiment but rather an experiment in disembodied ethical consciousness.[13] That is, *The Trace* is not a technological experiment in bodily extension but rather a psychic experiment into the necessary conditions for agency in a telepresent environment. Put differently, *The Trace* tests the experience of distributed agency in its social component.

From this perspective, though, a reduction is nonetheless performed in the piece: rather than dissolving the participants' embodiments into interchangeable and unspecified bits of data, *The Trace* seemingly reduces their mode of relating to a normative unilateral narrative. That is, *The*

Trace translates the participants' complex intersubjective relations into a single mode of relating: the desire to tele-embody (or not). Though other actions may—and necessarily do—take place in the space, the conditions of their registration (as agential activities) frame them in relation to this narrative. Moreover, *The Trace*'s tele-embodiment narrative not only imposes a unilateral and normative relation but also a teleological one because it marks a clear point of completion in the piece. From this, two points obtain: first, *The Trace*'s digital translation of the participants' ethical consciousness is reductive because it is bodily underspecified, in that it does not register those elements of its performance that exceed its dominant narrative (except as excess); second, it is reductive because it is relationally underspecified—the tele-embodiment narrative is a dramatically simplistic account of desire. In short, the work's shift from the territory of embodiment to that of relation displaces the violence of computation that Hayles cites, rather than eliminating it.[14]

My purpose in insisting on the "reductive-ness" of this aspect of *The Trace* is not to criticize the work but rather to highlight the fact that it is demonstrative of a transition from an aesthetics of technical accomplishment to one of a technologically disciplined subject. Read in this way, *The Trace* can be rethought as describing the limits of the discourse of embodiment with respect to distributed agency and its concomitant relational ontology. Put simply, the piece performs an intensification of the problematic of the subject that flips it from being an investigation of the subject's embodiment (a dominant concern of telepresence art in the early 1990s) to one of the ethical relations available to and between (embodied) subjects in the digital environments that intermediate contemporary technoculture. This, as discussed in chapter 3, is precisely Hayles's project. However, as I showed in chapter 3, Hayles employs an a priori set of ethical distinctions to move outward toward an account of digital relations; in contrast, *The Trace* gestures toward the inverse direction, utilizing the relational constraints imposed by its digital environment to suggest a particular ethical project.

SIMPLY PUT, *The Trace* demands a theoretical framework that can connect to both the participants' relations to one another and to the distributed agency and recursive feedback loops of their digital mediation—or, put differently, a posthumanist approach. I will argue that this framework might productively connect to the constitutive subjective ambivalence that Judith Butler (supplementing Foucault and Lacan) has theorized under the rubric of power, because many of the issues that *The Trace* raises in a digital context are taken up by Butler in their analog aspect. Taking account of how paradox structures the debate of subjectivity—always culminating "in displays of ambivalence"[15]—Butler's account offers a model from which to address the constitutive ambivalence of binary code, suggesting that code's very constraints might productively perform a critique of the unilateral narratives that structure it.

In this respect, Butler's rethinking of Freud merits attention. While Butler agrees with Freud that the mode of desire is always melancholic, she nonetheless argues that the constitutive foreclosure of this melancholia is not that of incest (as Freud argues). Instead, for Butler, homosexual desire is the primary foreclosure in subject formation (existing prior to that of incest) so that the constitutive melancholia of subjectivity is the active absence (or disavowal) of this desire, which in turn attests to the existence of that which is foreclosed. That is, in its very disavowal (in its radical foreclosure), homosexual desire is sustained as a structural necessity in the (always repeating, citational) process of subject formation.[16]

Thus, in its most basic formation, the subject is melancholic, and melancholia, in *its* most basic formation, is properly understood as an "internalization" of the Other. That is, whereas grief represents an externalization of the self into a social field of relationality (through the process of recognizing that a part of oneself has been lost in the loss that is the object of grief), melancholia (as foreclosure) is unable to recognize the object of loss such that loss itself, in its most unknowable dimension, is taken in as the subject. In this sense, the subject is always a radically ambiguous composition of self and Other, just as the porous boundaries of the social are always penetrated by (and penetrate) those of the individual. In a basic equation, then, the ambiguity of the subject and the relationality of the social are correlative for Butler, and their mode of relation is that of vulnerability (because the intelligibility of each is at the mercy of the other).

From this, Butler's political wager lies in the belief that to think of melancholic subjectivity as deeply embedded (in and as us) is to open the possibility of positivizing the aporetic ambiguity of our subjectivity in the form of political agency. That is, Butler does not think of agency as the expression of an agential subject but rather considers it as the mode of avowing a relation of vulnerability that is constitutive of agency. And isn't this precisely the scenario that is implicit in Hayles's construal of posthumanism, where no a priori individuation of wills is possible?[17] What Butler articulates, then, is the necessity of rejecting notions of subjectivity that reference preexisting identities in favor of those that operate a perpetual and socially contextualized process of identity formation.

To this end, Butler positions unilateral narratives of political dominance as treatable symptoms of melancholic subjectivity. In contemporary North American political culture, for example, she argues that melancholia has tended to give rise to narratives that foreclose the possibility of grieving certain human lives.[18] In response to this melancholia, grieving is an attempt to suture the gap of ambiguity that is constitutive of our subjectivity; though never entirely successful, this attempt opens (for Butler) the possibility of an ethical political agency. In short, although we can never fully avow the lost object of homosexual desire (which would be to obliterate our subjectivity), the ethical–political agency opened through the concept of ambiguous, ambivalent, and vulnerable relation traverses the fantasy (in Žižek's sense) of an agency composed of unilateral narratives.

This is precisely the line of critique that Butler levels at Lacan in *Antigone's Claim,* the central argument of which is that, in addition to functioning to prohibit sexual exchange among kin relations, the incest taboo "has also been mobilized to *establish* certain forms of kinship as the only intelligible and livable ones."[19] That is, even when it is only operative as a normative structure, a unilateral narrative of kinship—which, according to Butler, underwrites Lacan's thought—carries within itself its own menace. Specifically, this menace is implicit in the deviation that is constitutive of the reiteration of the Law that is necessary for kinship's structural operation. For Butler, then, the question is what this unilateral narrative of kinship forecloses, a question that she answers with a series of further questions pertaining to the contemporary status of structural figures of kinship.[20]

Butler's elaboration of this implicit menace of deviation recalls Derrida, whose famous critique of structure through the notion of supplementarity includes the argument that repetition always requires deviation.

This connection to Derrida is important here because Butler's analysis might be criticized for a lack of clarity with respect to what the criteria are for deeming a "structure" structural. That is, at what point does a narrative figure become a necessary site of deconstructive analysis? In this line, while Butler convincingly critiques Lacan for positing the heterosexual family as a unilateral narrative of kinship, could not the same criticism be leveled at the forms of sexuality, gender, and desire as they are found in Butler?[21] Although it is clear that Butler thinks of these categories in the most fluid terms possible, they remain—in a categorical sense—uniformly and unilaterally operative, which is really just to reappropriate Butler's interrogation of Lacan to ask, if gender, sexuality, and desire always allow for multiple definitions of themselves, where is this multiplicity (intelligibly) registered, and what sorts of meaning does the normative matrix of this site itself foreclose?

This is not to oversimplify Butler's position: to a certain extent, this line of critique is exactly what she aims to address through her emphasis on ambivalence.[22] Because Butler's subject is always in flux, its menace—paradoxically—both conditions and threatens it. To return to The Trace, then, we might say that the piece as such consists in performing this ambivalent flux: the participants' activities are not so much acts of "self-expression" as they are testaments to the ambivalence of The Trace's—and, more generally, virtual reality's—codes of interaction. In this, the piece constructs an ethical model through which the participants' intelligibility is revealed to be vulnerable, both to one another and to the way that desire is constructed in and by the piece.[23] It isn't just that the parameters of the piece normatively limit how the participants relate to one another; in addition, the piece itself is dependent on the excesses of the participants' flesh: the piece doesn't really make sense as a piece without the push-and-pull ambivalence of this codependence. The explicit process of normative reduction that The Trace performs is just as integral to the piece as the unilateral narrative of desire that this reduction produces.

In sum, two dynamics are at work in The Trace, and it is crucial to disarticulate them: on one hand, the participants' interactions, and on the other, how the piece, from a quite different vantage, stages those interactions.[24] That is, The Trace registers the ambivalence of digitality's claim to media convergence—of the process through which information becomes disembodied—by always simultaneously signifying the opposite of what is claimed: the participants act virtually within a reductive teleology of (simplistic) desire, for example, but this multimedia convergence also registers as a multitude of disjunctions in their flesh-world bodies (because they can't remain in control of all the various media). This is not to say that *The Trace* is somehow proof positive of Butler's account of melancholic subjectivity but rather that the subjective ambivalence performed in the piece means that the territory on which *The Trace* registers itself, makes itself intelligible *as a piece,* is that of Butler's melancholic subject. In registering ambivalence as its ground, then, *The Trace* suggests an ethics of ambivalence, which is to say that it suggests ambivalence—in this case, figured as active deferment—as a ground for agential activities.

Indeed, this ambivalence is also registered in the boundaries of *The Trace*: while the piece proper exists within the two digital stations, it also spills beyond this containment through the way that it negotiates the participants' entrance. Specifically, while prospective participants wait in line, often for quite a long time, they are able to view monitors showing the current participants as well as their statistical data. In this, the participants' subjective experience within the piece is also an objective experience for those who are waiting: because every participant must wait in line before entering, this objectification of the piece's digital reality penetrates the subjective experience of it while it is taking place. Of equal importance to the monitors, then, is that the line to enter the piece institutes a social scene prior to (and after) that of the participants' immersion. If *The Trace* is an experiment in relation, this waiting area acts as a liminal space between flesh reality and the piece's virtually extended reality. In turn, this suggests that the piece requires a boundary to articulate its version of virtual reality but simultaneously deconstructs this boundary through the interpenetration of subjective and objective experience that it occasions. In this sense, *The Trace* sits precariously on the point "where technologically mediated relations become . . . social relations."[25]

This immediately brings to mind an objection: doesn't reading the piece in this way neglect the fact that the participants have chosen to participate in the first place? That is, can a participatory artwork really be considered a valid testing ground for theories of subjectivity, or is this possibility precluded by the participants having had the option of not

FIGURE 5. Prospective participants line up before entering *The Trace* in Montreal. Photograph by Antimodular.

entering the piece in the first place? This is a valid argument but one that points equally back to the account of agency offered by Butler. That is, meaning itself remains structural in Butler, even as particular meanings are revealed to be contingent. Put differently, a certain conflation of reason and cognition remains the condition of possibility for registering Butler's critique,[26] which is really just to insist on a basic point: the moment that the possibility of positively advocating for an ethics (even one of ambivalence) emerges is, and will always be, simultaneously the moment that the menace that is internal to meaning also emerges. That is, the emergence of ethics is the ultimately arbitrary constitution of a restricted perspectival field, where the premises and consequences of an action are calculable to precisely the extent that they are meaningful: if meaning is present, then it is part of a normative operation. The question that pertains to *The Trace,* then, is whether it succeeds (alongside Butler) in performing ambivalence in its radical character as the end of determination. The answer to this question is not clear, which is to say that Butler's ethics of ambivalence, which is presented here as the ethics of *The Trace,* is itself ambivalent. The answer to the question is a field of other questions.

THE "FIELD OF OTHER QUESTIONS" that emerges in Butler points to a broader critique of her argument that will ultimately lead our discussion more securely back to *The Trace.* Specifically, consider Butler's substitution, in Freud's schema, of the foreclosure of homosexual desire for the incest prohibition. When Butler notes, in a footnote, that "presumably, sexuality must [also] be trained away from things, animals, . . . and narcissistic attachments of various kinds,"[27] we can understand that "homosexual" functions in her writing as a sort of stand-in for the field of desire itself: it isn't just homosexual desire that is foreclosed by the incest prohibition but all nonhetero (and indeed nonhuman) desiring. Paradoxically, then, to register desire is to institute a narrative that forecloses desiring, because the latter cannot be indexed to anything outside of itself. As a result, the lost *object* of Freudian melancholy (and the Lacanian subject) is replaced—in Butler's schema—not by a lost object of homosexual desire (as she usually claims) but by a lost *relationality* (or "relational field") of desire.

Again, it bears emphasis that this critique is simply a heightening of precisely what Butler is arguing: the shift from lost object to lost field that I am identifying can be understood as the shift from a subjectivity that operates relatively autonomously and in relation to the established norms of kinship to the radically relational subjectivity that Butler advocates and that has here been claimed as what is performed in *The Trace.* That is, in line with Butler, we might say that a shift from lost object to lost field signifies a transition from thinking desire as a problematic of the subject's relation to the social (where the two categories are distinct, but related) to thinking it as a mode of action, the *effect* of which is the (always repeating and thus changing) constitution of the subject and the social, of the performer and the performance. In this latter figuration, then, desire is not a force that seeks to establish relations but is rather the relations themselves understood as ongoing events (i.e., as desiring).

Thus the shift from object to field points to a further ambiguity in Butler's writing: if desire articulates the conditions for intersubjective relations, then does it not also act as a Law that conditions the operation of normativity? That is, just as a logic of sense conditions the registration of meaning, does it not also structure desire? If this is the case, then it follows that the normative force of Law does not spring from primal kinship relations but rather from a value-form of desire.[28] That is, desire is predicated—paradoxically—on its own foreclosure, on a meaning that both grounds it and follows from it.

The question, then, is to what extent desire remains an active force of ethical agency (which Butler's characterization of melancholy requires it to be) if it is always-already also the normative Law that it opposes. That is, does Butler's emphasis on the constitutive deviation of repetition (with respect to normative discourse) still hold if the Law and its deviating repetition both remain constrained by an injunction to mean? Put differently, does Butler succeed in articulating a form of desiring that does not relate to a determined form of desire? To a value of desire (in Nietzsche's sense)? Or, thought from the other direction, does this perspective offer a model of agency that ultimately has no way of registering itself as such? Obviously, this is a line of questioning that is not entirely unattended in Butler, but it is also not resolved; the question remains as to whether Butler's ambivalence is the radical potential for agency that she hopes it is or whether it is simply the ambivalence of a construction of meaning that is relentlessly haunted by meaninglessness. By performing this problematic, *The Trace* puts forth these questions—as questions—as an ethics of the hypermediated relations and distributed causalities that characterize Hayles's humanist construction of technological posthumanism. That is, rather than presuming an ethical framework for its performance of technological posthumanism, the "content" of *The Trace*'s performance is precisely—and paradoxically—a contestation of this context.

FROM HERE, we might understand *The Trace* as a piece that ambivalently performs the "turn" that Butler identifies as a condition of the subject's social intelligibility.[29] However, the multisensory telepresence of *The Trace*'s participants—achieved through the characteristic simultaneity of digital technology—renders this performance in a way that calls the "conditionness" of this constraint into question. That is, while the (apparent) content of *The Trace* is clearly a reduction of the participants' subjective ambivalence, the way that this reduction is acted performs a critique of unilateral narratives of relation and of the symbolic form of relation itself.

In this respect, *The Trace* concretely demonstrates the more general challenge that digital telepresence presents to ethical schemes grounded in anthropocentric constructions of relationality. As Virilio famously noted when the technology was still in its relatively nascent stage, by generating a perspective from beyond the limitations of customary conceptions of proximity, telepresence

> disposes of the very notion . . . of touch . . . [to seriously upset] not only the distinction between "actual" and "virtual" . . . [but also] the very reality of near and far, thus questioning our own presence here and now . . . [and seriously affecting] the conditions of necessity for direct sensual sensory experience.[30]

If conventional notions of "touch" are thus obsolesced, *The Trace* reminds us of McLuhan's claim that precisely such obsolescence is "the cradle of art,"[31] which is to say, an important condition for probing a future where tactility and relationality are not the exclusive and de facto province of an uninterrogated and categorical "human."

All of this leads us back to Hayles's nisus for a posthuman ethics. Listening to Butler, *The Trace* suggests a posthuman that is constrained by meaning into a doubled relation with ethics: on one hand, *The Trace* simply reiterates the paradoxical human subjectivities that Butler identifies; on the other hand, though, in staging this paradox as a medial disjunction, *The Trace* suggests technological posthumanism as a process in which meaning is superseded by relation. In the first instance, ethics enters the scene (alongside Hayles) as a corrective to the dematerializing tendencies of computation but is thereby positioned beyond the pale of posthuman discourse. In the second case, the very possibility of a posthuman ethics is contested, but in this contestation, an opportunity is opened to specify what exactly ethics might consist of in our present—hypermediated—historical moment. In the ambivalent relation of the two approaches, then, is the ethical problematic that *The Trace* performs so that—ultimately—we might say that *The Trace*'s wager lies in the performative repetition that sustains this crux *as crux,* thereby maintaining its active relation to ambiguity. Put differently, *The Trace* opens a paradoxical space in which posthuman ethics might be thought (and performed), insisting on the undecidability of any particular schematic of posthuman ethics, while simultaneously demonstrating that we—as posthumans—nonetheless continuously decide.

PART III

5 FROM AFFECT TO AFFECTIVITY: MARK B. N. HANSEN'S ORGANISMIC POSTHUMANISM

> The new man is the old man in new situations; in particular, he is the particular old man best suited for the new situations.
>
> BERTOLT BRECHT, AS QUOTED BY HAL FOSTER IN "HOW TO SURVIVE CIVILIZATION"

> The potential of deconstructive analysis lies not in the simple recognition of the inevitability of exclusions, but in insisting upon accountability for the particular exclusions that are enacted and in taking up the responsibility to perpetually contest and rework the boundaries.
>
> KAREN BARAD, "GETTING REAL"

This chapter examines Mark B. N. Hansen's organismic construction of technological posthumanism, particularly as it is presented in his three key texts, *Embodying Technesis, New Philosophy for New Media,* and *Bodies in Code.*[1] While the subject matter of these texts frequently overlaps, the primary focus of each nonetheless unfolds distinct components of Hansen's expanding argument for a view of the technological posthuman that simultaneously avows the radical (as opposed to the relative) exteriority of technology and the primacy of embodiment: *Embodying Technesis* unpacks the "putting-into-discourse" of technology as it has taken place across otherwise divergent strains of contemporary theory, showing how this action disavows technology's material specificity; *New Philosophy for New Media* insists that there is an affective topology of human perception that is fundamentally tactile, encompassing vision such that the latter relates to a primordial "haptics" that grounds it; and *Bodies in Code* synthesizes these perspectives, insisting more strongly than either previous book that technology is bound up in human embodiment itself and that we can thus best understand human technogenesis through a rereading of the operational perspective invoked by certain theorists of autopoiesis. Collectively, then, Hansen's oeuvre seeks to give "a robust account of technology in its irreducible materiality that exists beyond discourse and representation,"[2] while also evincing an "'originary' coupling of the human and the technical" that grounds experience as such and that "can only be known through its effects."[3]

Put simply, I argue in this chapter that Hansen's perspective is ultimately haunted by the very representational logic that it moves against. In this, I do not repudiate Hansen's argument as such but rather reject one of its central underlying implications, namely, that the extradiscursive materiality of technology might be accessed, linguistically, without attaching a meaning to it *that is fundamentally foreign to this materiality*. Such access, then, generates a bias that is naturalized through the notion of technology per se because the latter masks its contingency. In particular, I argue that this bias is already nascent in *Embodying Technesis* as the concept of *technesis* itself and that it reaches full maturity in the organismic perspective that grounds the title term of *Bodies in Code*. To this end, the chapter begins with an examination of technesis as it is initially developed by Hansen, demonstrating the necessity from which it sprang, the contribution that Hansen's reading makes, and its ultimate limitations. From here, the chapter articulates Hansen's argument for an affective topology of the senses, corroborating the increased importance of digital technologies in this perspective through a brief comparison of Robert Lazzarini's *skulls* (as read by Hansen) and my own piece *Sound*. Finally, this comparison pivots the chapter toward a critical analysis of Hansen's robust account of primary tactility, an analysis that concludes the chapter by agreeing with most of Hansen's assertions but also by insisting that they remain within the (representational) logic of language. In closing, I argue that what is accomplished by Hansen's putting-into-discourse of technesis is, paradoxically, a restaging of the constitutive ambivalence of deconstruction that shows the latter to be a promising premise for specifying the relation between humans and technology.

EMBODYING TECHNESIS came into print in the midst of what Hansen identifies as a shift in science and technology studies, moving toward a paradigm that gives precedence to concrete experiential effects over abstract theoretical significance. This change was necessitated by the dramatic disjunction existing between technologists and theorists of technology, aptly summarized by N. Katherine Hayles:

> From the point of view of the technologist, the critical theorist knows nothing about the workings of technology and demonstrates it by talking in terms so abstract that material objects are vaporized into mere words; from the point of view of the theorist, the technologist is so stuck on nuts and bolts that he remains oblivious of the fact that these objects are never simply present in themselves, being always already enframed by cultural assumptions and mediated by discursive practices.[4]

Thus, with the aim of conjoining material practice with theoretical sophistication, the conventional "top-down 'diffusion' model" of analysis has been increasingly replaced by translative approaches that attend to the minutiae of everyday activity.[5] As emblematic of this perspective, Hansen cites Isaac Asimov's famous suggestion that the lived experience of modern man has been more dramatically shaped by practical inventions, such as the automobile, than it has by more (discursively) consequential shifts in scientific knowledge, such as Einstein's theory of relativity. What is most important to attend to, in this view, are thus not the major events, ruptures, or characteristics that are readily available as objective knowledge in a historical macronarrative but rather the complex relations that contextualize these events and compose the ground for the narrative itself.[6]

In principle, this is a shift that Hansen agrees with wholeheartedly so that the task he finds currently at hand is to specify, precisely, the ways that everyday technologies profoundly impact our lives, rather than simply insisting that they do so. However, in the case of technology, this is not so simply accomplished because (for Hansen) "technologies *underlie* and *inform* our basic 'ways of seeing' the world":[7] the ways in which our daily activities presuppose technology function to foreclose the possibility of analyzing them in isolation. As a result, Hansen insists that "despite its irreducible concreteness, technology's experiential impact must accordingly be considered to be first and foremost indirect and holist."[8] In short, to the extent that technologies structure perception, they also elude it, resulting in the tremendous difficulty—perhaps even the impossibility—of thinking technology in its own terms.

For Hansen, this evasion not only explains the tendency to think technology abstractly but also (in a related way) leads to what he calls the "culturalist assimilation of technology."[9] In the case of the latter, Hansen insists that all too frequently, the (in his view, true) observation that technologies cannot exist outside social systems is conflated with the (in his view, false) claim that they can "be captured by the interpretive tools germane to such systems."[10] Thus Hansen does not take issue with the culturalist position that understands technology to be part of a complex social network, but he is resolute that technology is also more than this, and radically so.

Indeed, it is precisely this radicality that is at stake in Hansen's overall project, as one of the motivating factors in his writing is the claim that twentieth-century theorists—"from Freud and Heidegger to Lacan, Derrida, and Deleuze and Guattari"[11]—relativize the exteriority of technology, subordinating it to the more central overriding theoretical purposes of each of their intellectual projects. In this, these theorists interpret the "host of concrete materializations through which technologies impact our practices . . . according to 'logics' that are strongly posthermeneutical," ignoring the fact that technology influences our embodied lives "at a level below the 'threshold' of representation."[12] In so doing, Hansen argues that technology is rendered an object of knowledge and is thus exchangeable for other objects within this logic of representation. Simply put, so long as technology is taken to be grounded in representation, its (for Hansen, innate) potential to intervene from outside of this system is effaced. In the context of manifold cultural, discursive, and linguistic "turns," Hansen echoes Karen Barad's observation that "there is an important sense in which the only thing that does not seem to matter anymore is matter."[13]

Hansen uses the term *technesis* to account for this representationalist reduction, defining it first as the "putting-into-discourse" of technology.[14] In this, Hansen traces an explicit link to Alice Jardine's analyses of *gynesis,* where she points out that the entrance of "woman" into discourse frequently yields situations where "woman" may "become intrinsic to entire conceptual systems [that are nonetheless not] 'about' women, much less feminism."[15] Though both *technesis* and *gynesis* thus indicate an abstraction into language, however, Hansen emphasizes that the two reductions nonetheless perform different strategies: whereas gynesis perpetuates a reductive use of "woman" in service of expanding our understanding of textuality, technesis valorizes the text as a model for understanding technology so that its primary function is to translate technology into a textual analog. Thus, although technesis employs a logic that is similar to gynesis (and other structural modes of analysis), the reduction that it enacts is, for Hansen, far more violent: *technesis* not only reduces technology by underspecifying it but it actually alters the materiality of technology by respecifying it differently. Simply put, "technology itself" disappears in technesis, except in the form of alibi. In Hansen's reading, the representation of technology through technesis bears about as much similarity to the actual material specificity of technology as a bank robber wearing a Richard Nixon mask bears to the former president: in both cases, the contents of the mask (i.e., technology and Nixon's facial features) are subjugated to the logic of the events in which they are staged (i.e., language and a bank robbery).

In *Embodying Technesis,* Hansen offers numerous ways in which this representationalist reduction is enacted, emphasizing the particular disciplinary agenda that is put to work in each case. For example, he points out that systems theory functions "by bracketing out all constraint the real might impose [on a system]"[16] such that materiality "receives a purely abstract determination as that which resists translation *per se.*"[17] Similarly, he accuses cultural studies of being generally "restricted to the effects [technology] has on our capacity to constitute ourselves as subjects, to represent ourselves to ourselves, . . . [so that] whatever exteriority is thereby broached [can only be] a relative exteriority, a point of resistance *internal* to the representational space of thought."[18] In both cases, technology is abstracted into the disciplinary logic that discursively precedes it, ignoring the fact (for Hansen) that technology exists in its own right, within a stratum of reality that is fundamentally prediscursive.

Importantly, then, both these examples (as well as others) flow from Hansen's broader, and more scathing, critique of Derrida's deconstructive grammatology in *Embodying Technesis,* particularly the former's claim—in Hansen's view—for language as "the exclusive or privileged faculty of experience."[19] Via a critique of grammatology, then, Hansen mounts his thesis that technesis employs technology as an abstract cipher for an internal otherness that, while constitutive of texts, does not pertain to technology per se. In turn, this criticism feeds back to the other examples that he offers, demonstrating how, in each case, linguistic logocentrism—despite manifold claims to the contrary—continues to dominate the various ways in which technology has been engaged within (at least Western) philosophical discourse.

However, this critique of Derrida merits further consideration. Hansen's specific argument notes that within Derrida's hugely influential work, technology is consistently restricted to a "model of the text as machine."[20] In particular, he sees in Derrida (as in other twentieth-century poststructuralist theorists) an inversion of the machine metaphor as it had developed since at least the seventeenth century (predominantly in the form of the clock), wherein machines were "consistently employed as a heuristic for conceptualizing what is proper to the human."[21] Through this inversion, a machinic ontology of textuality is produced in which "language assumes the role of a machine that runs independently of the phenomenal and rhetorical categories governing lived experience"[22] such that "for the first time, the machine is actually deployed as a metaphor *for technology itself.*"[23] In this, Hansen argues, technology is domesticated by Derridean textuality such that it is dependent on the movement of *différance* to the same extent that writing (in the restricted sense) is. As such, it is not a site of inquiry in its own right but is rather restricted to a "doubly derivative status" that simply lends support to "the totalizing grasp of [Derrida's] ontology of *différance.*"[24]

The merit of this critique of Derrida is that it articulates, constatively and performatively, the tremendous difficulty of the task that Hansen has set himself in trying to account for technology "in its own terms." Its drawback, though, is that it rests on a fundamental misreading of Derrida's grammatology. Emphatically, Derrida does not idealistically reduce everything to language when he insists (in *Of Grammatology*) that "there is no outside-text":[25] as has frequently been noted, the popularized translation of "Il n'y a pas de hors-texte" as "there is nothing outside the text" is both incorrect and misleading in that it suggests an ontological understanding of language (where language would, in theory, be capable of total description). Instead, Derrida's point is that "what opens meaning and language is writing as the disappearance of natural presence,"[26] in the sense that language is a condition of legibility, even if it always renders its objects paradoxical and incomplete. Importantly, then, the term *open* indicates that Derrida does not suggest a claim for language as "the exclusive or privileged faculty of experience,"[27] as Hansen's argument requires. Instead, Derrida's formulation insists simply that insofar as there is meaning (i.e., signification, sense, an interpreted goal, etc.), this meaning is contingent on the instability and ambiguity inherent in language. Thus it is resolutely not Derrida's belief that language determines bodily experience per se but rather that it is the technology through which experience is registered as such, and that this registration reveals that both language and experience are always-already both present and absent in their relation. For example, my digestive system might well behave extralinguistically, but my experience of it as a functional operation (i.e., a system that digests my food) is linguistic; I can say neither that my digestive system preexists, for me, its meaning (which would be to say that it preexists itself) nor that the meaning of my digestive system preexists it (because I can only know it insofar as I can experience it).[28] For Derrida, the same is the case for technology.

Put simply, then, the object of Derrida's critique is not experience, as Hansen would have it, but rather our knowledge of it. That is, to the extent that the machine appears in Derrida as a metaphor for technology itself, it does so *as a metaphor,* which is to say, as a substitution for the impossible necessity of technology, for its paradoxical presence–absence. Indeed, this ambiguity is even exemplified in a passage from Derrida that Hansen cites in his argument. Consider the following:

> A certain sort of question about the meaning and origin of writing precedes, or at least merges with, a certain type of question about the meaning and origin of technics. That is why the notion of technique can never simply clarify the notion of writing.[29]

For Hansen, this passage exemplifies the machine's "foundation in *différance*"[30] for Derrida, a view that positions grammatology as an identifiable ontology. However, this neglects the crucial hesitation in Derrida's writing: in his parenthetical "or at least merges with," Derrida gestures to the paradoxical temporality at the core of deconstruction, the radical impossibility of answering questions of "meaning and unknowability." In

Of Grammatology, this impossibility is figured as a trace (or *arche-trace*), a mark that "was never constituted except by a non-origin . . . which thus becomes the origin of the origin" and which cannot be marked by empiricism.[31] Clearly, then, Hansen's implication that Derrida submits technology to a structural logic only tells half the story and misses the explicitly strategic aspect of Derrida's writing: to the extent that Derrida positions writing as preceding technology, he does so keenly aware (as Spivak notes) that he is operating "according to the very vocabulary of the thing he delimits."[32] Indeed, this is why the *notion* of technique can never *simply* clarify the *notion* of writing; they are all bound together in a paradoxical linguistic logic (a notionality?) that always performs something supplementary to its claims.

Thus, to say that technology is reduced by grammatology is problematic on two counts: first, it ignores the fact that this statement is in complete agreement with grammatology, to the extent that it consists in the claim that technology can never be fully represented—can never be fully present—in language; second, and more egregiously, it assumes that such full presence is in fact possible, in the sense that language might reduce technology rather than ambiguously constitute it. Thus Hansen's claim that grammatology reduces the exteriority of technology rests on a translation of grammatology into a totalizing structural system of meaning, neglecting that all writing (in the broad sense that grammatology promotes) must "be capable of functioning in the radical absence of every empirically determined receiver in general."[33] Moreover, this reduction allows Hansen to naturalize technology outside of language, while nonetheless operating its force within it: if technology is truly radically exterior to language in the sense that Hansen indicates, how could it possibly interact with the linguistic paradigm that characterizes the type of meaning that Hansen requires?

Hansen is certainly aware of these quandaries but believes that they are symptomatic of a "culturally ingrained logocentrism that compels us . . . to translate technological materiality into discourse."[34] In contrast, then, Hansen turns to Bourdieu's account of mimeticism to develop a notion of a bodily hermeneutic that avoids "translation or delegation into language [by requiring us] to learn to use our mimetic bodily 'sense.'"[35] This "practical mimesis" takes place below the threshold of knowledge as it is staged in representation and thus yields embodied "knowledge" that can "only be experienced through mimetic reproduction, never through translation into language,"[36] a necessity that (Hansen notes) Bourdieu suggests is "particularly clear in non-literate societies, where inherited knowledge can only survive in the incorporated state."[37] For Hansen, then, developing this mimetic bodily sense is a promising method for unshackling our ability to experience our bodies from what he sees as the profoundly restrictive dependence on representation (and cognition) that grammatology imposes.

However, this raises the question, isn't cognition included in embodiment? That is, it is at least possible that the "mimetic faculty" that Hansen suggests already exists as cognition, as the body's way of representing itself to itself. Indeed, Hansen gestures toward this in his frequent invocation in *Bodies in Code* of the later Merleau-Ponty's notion of a primordial *écart,* an internal fissure that is "the fundamental dehiscence that explains the body's need for the world"[38] and that acts as a transducer "between embodiment and specularity."[39] However, whereas Merleau-Ponty mobilizes the *écart* in embodiment to describe the flesh as an "intertwining [and reversibility] of the sensate and the sensible . . . [that] directly problematizes the concept of intentionality,"[40] Hansen maintains that the *écart* occupies the "cusp between the biological and the psychic *prior to their actual differentiation.*"[41] As a result, embodied alterity is, for Hansen, "simply a primary condition of the being of the body"[42] so that Merleau-Ponty's understanding of the flesh is folded back into Hansen's concept of the primordial body.

Thus Hansen's objection to the suggestion that cognition might constitute a "mimetic faculty" might be that embodiment remains subjected to representation in this reading, but such a claim would neglect (as shown earlier) the profound avowal of that which is erased in this same understanding of representation. Moreover, to the extent that grammatology recognizes a materiality of language, does this not indicate the kind of sensory retraining that he is suggesting? It is a slippery point, but I would suggest that this is so not only because—as Hansen insists—there is a logic of embodiment that is external to language (which may be true, though to what extent the two might interact remains in question) but also because identifying the problematic of embodied knowledge as distinct from that of language depends on a (provisional, at least) separation of mind and body that has by now been so thoroughly and routinely deconstructed that—for most contemporary readers—the separation of the two cannot be sustained. In this sense, Hansen falls into the trap that Hayles describes in her differentiation of embodiment from the body in that he (implicitly) posits the body as a primordial blank slate abstracted from any context. That is, Hansen produces a normative "body" that is not connected to any particular embodiment, thereby taking the body as a given.[43]

Thus, to the extent that Hansen critiques the (reductive) identities produced by representation, he is perfectly in agreement with a deconstructive approach. Where he differs is in the belief that we can intentionally develop faculties that are—and forever remain—fundamentally foreign to intention but that are still meaningful. If we were to really take seriously his call to develop a bodily sense that is not subject to language, then, we would have to do so without having set out, intentionally, to accomplish this. Not only is this impossible, but it is an impossibility that is structured according to the very paradoxical logic of language that Derrida's grammatology writes.[44]

Of course, Hansen is not unaware of this paradox. Indeed, the final pages of *Embodying Technesis* turn to Benjamin as a means of pointing "beyond the impasse of *technesis* [by] refusing to collapse the technological real into representation and by linking it to embodiment."[45] Again, though, this reinforces more than undermines Derrida's account.[46] Moreover, how precisely we can "make *sense* of technology's diffuse, amorphous corporeal impact without filtering it through language" remains decidedly underdetermined in Hansen's text.[47] Or rather, perhaps it is performatively overdetermined, which is to say that *Embodying Technesis* performs a leap of faith—a leap into impossibility—that it cannot avow. What Hansen accomplishes, then, might be not so much a movement beyond language to technology as a technological restaging of the dramatic and paradoxical ambivalence that language performs.[48]

IF HANSEN'S NOTION of technesis does not convincingly refute—or even sidestep—the textual logic of Derrida's grammatology, it nonetheless remains a powerful analytical tool for specifying the interaction of language, embodiment, and technology. For Hansen, this interaction *is embodiment* in the broad sense, which is to say that he insists that "any system involving 'information' requires an interpreter, and that interpreter is the material human body grounded in the wetware of our sensorimotor systems."[49] As a result, there is (for Hansen, in Barad's words) a strong "sense in which historical forces are always already biological."[50] Importantly, then, Hansen mobilizes an affective understanding of embodiment that accords it not only ontological priority (as suggested in the preceding statement) but also perceptual precedence. For Hansen, the body exists in its affective dimension prior to its articulation in the senses; furthermore, it actually serves as the frame through which these senses come to be understood as such. As Lenoir aptly summarizes, in Hansen's account, "the affective body is the 'glue' that underpins consciousness and connects it with sub-perceptual sensorimotor processes."[51]

Hansen's specific articulation of this presubjective affectivity, especially in the mature form that it takes in *Bodies in Code,* is considered in greater detail later. Here I would like to emphasize the relation of vision to embodiment that this perspective entails and particularly how this relation is revealed in and through contemporary technologies. Understanding Hansen's account of vision will help to contextualize his views on perception in general, which in turn will elucidate the precise sense in which the body, for Hansen, exists as a unified field that precedes—logically, if not chronologically or developmentally—the differentiation of the senses.[52] Simply put, Hansen understands vision to be a figuration of embodied process rather than an abstract power that informs it. In this, vision is not (strictly speaking) a medium but is instead a technological extension of the medium of embodiment.

Hansen illustrates this through an analysis of Robert Lazzarini's installation *skulls* (2000). First exhibited at the Whitney Museum of Modern Art, *skulls* consists of four sculpted skulls, each mounted at approximately eye level in the center of a different wall of a small, well-lit, clean, and bright four-walled room.[53] As described in the exhibition documentation, Lazzarini created each of the skulls by digitizing an actual skull into a digital three-dimensional geometry, where it was subsequently subjected to compound mathematical distortions. The results of these distortions were forwarded to a rapid prototyping machine that output a plastic prototype that served as a production model from which a rubber mold was made. A final version of the skulls was then cast in bonded bone: a mixture of resin, bone, and pigment.[54]

FIGURE 6. Robert Lazzarini, *skulls* (Whitney Museum of American Art, New York), 2000.

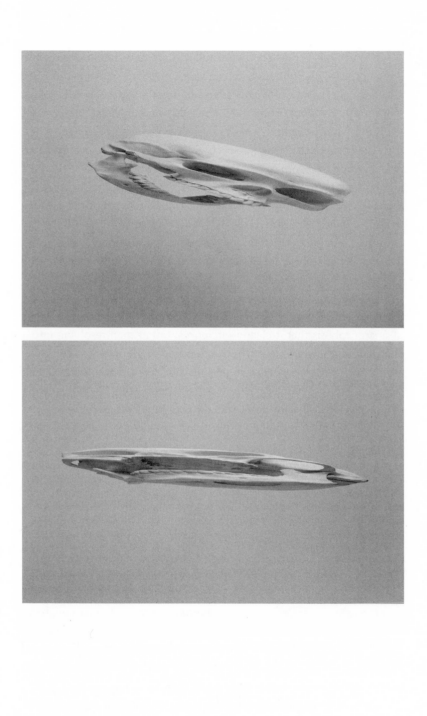

In effect, *skulls* translates literal and digital three-dimensionality into one another. This conflation "creates an unsettling spatial paradox [in which the skulls] appear to expand and contract as [the viewer's] vantage point shifts, suggesting something both static and moving."[55] As a result, *skulls* can never be entirely resolved in perspectival space in two senses: first, the "proper" viewing perspective is different for all four skulls arrayed around the room so that the visual solidity of one coincides with a liminal position in the others; second, and more important, because the skulls are modeled using mathematical—rather than sculptural—distortion techniques, the depth of the individual sculptural objects "interferes with the illusionary resolution of perspectival distortion."[56] As a result, the "perceptual experience of the work yields an oscillation or leveling of the figure-ground distinction."[57]

For Hansen, this ambivalence forecloses the possibility of visually mastering the work. In this, he notes that *skulls* differs from the icons of anamorphosis that it clearly references: rather than resolving into a "normal" image when viewed from a particular angle, it instead confronts the viewer "with the projection of a warped space that refuses to map onto her habitual [visually based] spatial schematizing."[58] In this sense, although *skulls* is visually perceived, the content of this perception is not visual per se in that it displays a logic that is fundamentally at odds with that of vision.

In fact, Hansen argues that what is seen in *skulls* is not even really a perception at all in the conventional sense but rather the disjunction that exists between human perception and digital technologies. In this respect, we don't so much *see* the installation as we do *feel*—affectively—the inaccessibility of its digital material to our senses. That is, *skulls* presents a "realignment of human experience *from* the visual register of perception (be it in an "optical" or "haptic" mode) to a properly bodily register of affectivity in which vision, losing its long-standing predominance, becomes a mere trigger for a nonvisual haptic apprehension."[59] Thus, although the installation does not afford "a direct apprehension of an alien space that *is* digital, [it does yield a] bodily apprehension of just how radically alien the formal field of the computer is from the perspective of the phenomenal modes of embodied spatial experience."[60] In so doing, *skulls* manifests (for Hansen) the radical exteriority of digital technology to the logic of vision but also how visuality (which includes, for Hansen, figurations of the other senses) is shaped in terms of "more visceral" bodily elements.

Importantly, then, Lazzarini's warped skulls institute a warped space that is sensibly felt by the viewer, even if only as a disjunction between digital materiality and the physical laws to which we are accustomed. The point, then, is that "the body continues to be the active framer of the image, even in a digital regime."[61] That is, Hansen argues that the spatial disjunction experienced in *skulls* is situated in and as the viewer: rather than being a gap between internal and external sensation, this disjunction is an interval within the viewer's body, a "bodily spacing" that indicates the precedence of an embodied organism over its sensory perceptions. Put differently, by rendering the spatial activity of the participant's vision useless (because the participant is unable to delineate her distance from the skulls through a conventional geometric perspective), *skulls* initiates a shift to an "alternate mode of perception rooted in our bodily faculty of proprioception."[62] In this way, the work functions by catalyzing an embodied sense of the skulls' form (i.e., an "affective process of embodied form-giving"[63]) that creates a place within our bodies that, in its creation, gives us a "sense of the 'weirdness' of digital topology."[64]

For the time being, the key part of this reading of *skulls* in *New Philosophy for New Media* is that, for Hansen, it positions *skulls* as exemplary of the potential of media art: rather than citing digitality as an abstract placeholder figured through arbitrary metaphors (i.e., the various visual images that we have come to associate with cyberspace), *skulls* "presents us with actual artefacts from the digital realm"[65] that are not "channeled through the coordinates of an image designed for interface with (human) vision."[66] As a result, the installation acts as a "cipher or index of a process fundamentally heterogeneous to our constitutive perceptual ratios":[67] the experience of the piece takes place as an affective response in the body of the spectator—it takes place "as the production of place within the body"[68]—that correlates to the work's digital topology, while nonetheless remaining radically discontinuous with it. In the experience of *skulls* (and other exemplary works of media art), then, Hansen believes that we can bodily access a material specificity of technology—or rather, we can appreciate the radicality of a particular technology's inaccessibility—without reducing it to language.

It bears emphasizing that to the extent that this is the case, it is accomplished via a movement of vision into the body that Hansen describes under the auspices of a bodily affective topology. With this in mind, consider as a counterexample the piece *Sound* (2001), a piano performance that consists of a combination of piano preparations; extended techniques; and loud, low, and quickly repeated notes played with the piano's sustain pedal depressed.[69] First presented in the context of a contemporary music recital, *Sound* utilizes microphones that are placed inside the piano, routed through a computer featuring sound-processing software, and plugged into loudspeakers. Despite this setup, though, the computer is not powered on so that the physical sound of the piano is in no way altered by the digital technology that is present: the piece, then, is simply a piece for piano . . . and yet it isn't.

Somewhat predictably, audience members often hear effects generated by the microphone–speaker–computer setup, despite the fact that no such sounds are literally present. What is more remarkable, though, is that this tendency obtains even in the context of highly trained listeners, for whom the sound-world of the piece is by no means novel.[70] Moreover, these skilled listeners—and I am not using the term *skilled* ironically—have even, in a number of cases, been able to specify the particular effects that they have heard: audience members have attested to hearing various equalization filters (e.g., high-pass, shelving, and parametric), delays, and shifts in synthesized reverb and dynamics. In fact, I have anecdotally noticed that the more sophisticated a listener is—the more she is able to specify what she's heard in a given piece—the more likely she is to have had highly complex and specified hallucinations during *Sound*'s performance.[71]

Emphatically, *Sound* is not a sonic hoax because it does not ultimately deceive the listener's ears.[72] That is, the sounds that the listener identifies as digitally produced are nonetheless audible and even properly identifiable in the sense that the listener is able to specify (via the terminology of digital technology) the particular sounds that she is hearing.[73] Moreover, it is plausible that the digital technologies presented on-stage actually serve to increase the audibility of the nuances of the piece in the sense that they signal a mastery of timbre that is often neglected in acoustic instruments (i.e., they suggest that the piano might be deinstrumentalized as a fixed medium through which piano notation passes, pointing instead to a particularity that is singular to a given instance of piano playing, and they additionally reinforce the fact that each of the piano's pitches articulates multiple frequencies). The digital technologies presented on-stage in *Sound* are not so much a hoax, then, as they are a rubric or cipher for the sound-world of the piece.

What is of interest here, though, is how this rubric actually functions. The point is that, in effect, the technology (not) used in *Sound* is more familiar (i.e., more accessible to our senses) than the acoustic phenomena from which it was developed: we tend to intuit (technological) mediation more readily than the claim to authenticity implied in conventional instrumentality. In this sense, the piece employs a simulation in which representations—the sounds of particular effects—are detached from any necessary or particular acoustic provenance. As a result, it is not only the case that the digital technologies offer a visual point of entry into the sound-world of the piece but also that they offer a point of departure from the linear causality—the instrumentality—of the acoustic instrument present (i.e., the piano, as the sounding mechanism of an original score or musical intention). In a mixed-reality context such as that suggested by the technologies presented on-stage, then, the physics of the performance—the question of what, materially, causes a given sound event—subtly shifts into a kind of pataphysics in which an imaginary sonic reality precedes its cause.[74] Simply put, the computer functions as an alibi for limitless representational manipulation so that the listener accepts that the sound events are not completely constrained by a recognizable causality. That is, the computer creates an illusion of a certain type of music—live electronic manipulation of acoustic instruments—that presents itself as music without illusions.[75]

Indeed, this "limitlessness" is the flip side of the tendency of listeners to project specific technical manipulations onto the piece: listeners' (mis) recognition of the processing is frequently accompanied by questions about other aspects of the sound-world of *Sound*. Often, these questions come from a lack of second-order identification of the phantom processes: for example, a phantom EQ might be heard as one that is regularly deployed in the program MaxMSP, which would lead to questions about other (phantom) manipulations that the listener did not recognize as being conventional in that programming environment.[76] Thus a series of questions are asked (e.g., "Is another software program being run in parallel?" or "Was that accomplished by misappropriating such and such 'object' in Max?") that not only presume the event but also understand it to be a technical fait accompli. Rather than creating disjunction in the listener— that is, rather than causing her to question whether things really are as they seem—these questions are invariably couched not only in the certainty that they took place but moreover with the assuredness that they took place as identifiable, predictable, and repeatable technical manipulations within the computer.[77] With the presence of the computer, the de facto assumption is that anything that can be registered sonically by the listener can also be registered within the computer and that with this registration comes a potential for manipulation that is virtually unconstrained. As Paul Théberge (among others) has noted, "a computer is both a machine and a social relation";[78] in *Sound,* the machine–computer is extended through the technology of the piano, which empties it of its instrumentality to feed it back as a paradoxical relationality. *Sound,* then, is virtual music par excellence, where the "boundlessness" of virtuality is acted under the alibi of computer technology that it does not otherwise employ.

This is notable in the context of this chapter because it is the opposite of what Hansen finds in *skulls.* Whereas Hansen demonstrates that *skulls* renders techno-logic visceral because it cannot be represented by the senses, *Sound* offers a "pure representation" of technology that instead captures viscerality in its exchange. In *Sound,* a "boundlessness" of the virtual is acted precisely through the way it opens the senses to its pataphysics, to a willingness to accept seemingly unconstrained causality without giving up the notion of causality itself (i.e., without leaving the realm of conventional representation). Thus, if *skulls* catalyzes a (disjunctive) perception of the grounding of vision in presensory embodiment, *Sound* suggests that the reverse also takes place. That is, in *Sound,* the affective body partakes in a logic of (non)representation in which the body itself moves into the object of its senses, *feeling*—"presensorily"—its profound complicity with the noncausal causality of grammatology. Simply put, the grounded and constrained physics of the body is evacuated into a play of representation.

Indeed, this understanding is coded into Hansen's claim—discussed in further detail later—that virtuality resides in and as analog subjectivity, which includes as its corollary the disarticulation of virtuality from digital technologies. That is, although virtuality has an affinity with digitality, it nonetheless (for Hansen) "stretches back to the proto-origin of the human."[79] In this respect, then, virtuality denotes the possibility for things being other than what they are, for the complex reorganizations that constitute human technogenesis.[80] Put bluntly, the virtual is (for Hansen) the space into which technics externalizes the body.

The difference with *Sound,* though, is that the body proper is not externalized in the piece but is rather translated (or reembodied) into a visual representation.[81] Moreover, this process feeds back to reinsist that the initial body is subject to representation: to the extent that the listener feels the piece, she does so within the logic that is suggested by its representations rather than via a spacing that would ground this logic.[82] To the extent that the body is externalized into virtual space, then, it is also revealed as always-already virtual.[83]

This does not in any way undermine Hansen's reading of *skulls.* However, as was the case in his critique of Derrida, what *Sound* does demonstrate is the leap of faith that Hansen must make to move from his perceptually and cognitively based reading of *skulls* to his positing of the body as primordial (and thus precognitive and preconceptual). Without this leap, for example, we would be forced to ask why vision is not accorded the radicality that technology is. That is, if vision takes place at an ontological level that is different from the embodiment that grounds it, doesn't this mean that there is a part of vision, however small, that is not accounted for in embodiment? As such, doesn't Hansen reduce vision to the affective logic of the body that he expounds? If so, what is accomplished in this reduction?

WHAT I HAVE ATTEMPTED in this chapter thus far is a simple recontextualization of Hansen's thought: rather than accepting his cogent analyses as proof positive of the structure of subjectivity that he advocates, I have demonstrated that his writing necessarily performs a gap that its content covers over. As a result, I have suggested that rather than an affective body preceding the senses, as Hansen would have it, the two necessarily exist in tension with one another. However, in insisting on this tension, I am emphatically not suggesting that affectivity is antecedent to the senses either but am rather importuning the impossibility of establishing a hierarchy in their relation. Put differently, though Hansen's perspective cannot be unequivocally true in the way that he hopes, it nonetheless performs an important intervention into the potentially stagnant discourse of deconstruction: if contemporary theorists have become comfortable with the paradoxical play of presence and absence brought to the fore in deconstruction, Hansen's writing serves a decisive role in reintroducing embodiment as a site of discursive discomfort. Thus, if Derrida's work was necessitated by a historical context that presumed presence, Hansen writes in a time that has perhaps—at least with respect to the humanities as they exist in academic institutions—moved too far the other way.

With this in mind, I offer in what follows a closer consideration of Hansen's account of affectivity, presented in an effort to specify Hansen's organismic construction of technological posthumanism as the twin presence of the manifold absences that have been made manifest in humanities scholarship over the past forty years. Hansen's posthumanism is perhaps uniquely suited to articulate this perspective because, of the three strains of technological posthumanism considered in this book, his perspective is in a sense most avowedly humanist: whereas Hayles, as I demonstrated in chapter 3, *insinuates* humanist values into the ground of her claims for expanded understandings of embodiment, materiality, and meaning, Hansen restages these expansions within the internal spacing of presensory human affectivity. Accordingly, Hansen acknowledges that it is specifically human embodiment that "takes on a truly unprecedented responsibility [at this particularly crucial moment in our coevolution with technology]: the responsibility of constraining and thereby specifying the process through which information 'objects'—images, space, events—are actually generated."[84] This distinction between Hansen and Hayles is crucial because it relates directly to how value—broadly construed to include meaning, ethics, and politics—operates in the schema. For Hayles (and, to a lesser extent, for Butler) value appears as an injunction to mean so that the exciting possibility opened by new technologies is their affordance of new modes of *represented* individual agency: value is external in this reading, in the sense that it is a crucial motivating principle in Hayles's thought. For Hansen, in contrast, value is *produced* via the body's internal spacing so that what is exposed by virtual technologies is "the violence exerted on bodily life by generic categories of social intelligibility and the politics of recognition."[85] Thus, if Hayles's technological posthumanism ultimately reinscribes the human, Hansen's does so doubly, but with a key difference inhering in how they position their interventions in the discourse of technological posthumanism: whereas Hayles figures her thought as unwinding embodiment outward into intermediating machinic materialities (that themselves feed back into bodies), Hansen works in the inverse direction, folding new technologies—including new modes of perception—back into the primary tactility of embodied subjects. The point, then, is that Hansen is not interested in new technologies for their potential to reconfigure *what* value is but rather for their potential to reconfigure *how* value acts.[86] This shift, for Hansen, is the shift from affect to affectivity.

To grapple with this shift, it is first necessary to exfoliate the particular understanding of technology that is at the core of Hansen's thought and that exists in contradistinction to the notion of technesis discussed earlier. Fundamentally, and in accordance with both Dyens and Hayles (as well as numerous others), Hansen considers technology to be intimately bound up with human embodiment. Quite simply, Hansen argues that every technology exists in relation to human embodiment and speaks to "the body's role as an 'invariant,' a fundamental access onto the world."[87] Although technologies may "refunctionalize"[88] the body, then, they do not institute any ontological shift that would threaten phenomenological accounts of embodiment. Instead, Hansen understands technicity, in its broadest sense, to be a process of "exteriorization."

Importantly, though, Hansen does not consider this relation to exteriority to be something "merely added on to some 'natural' core of embodied life" but rather takes technicity to be "a constitutive dimension of embodiment from the start."[89] In this respect, Hansen echoes Butler in registering a "turn" at the core of our subjective relations, for Hansen hinging on technology: technologies extend our interface with the environment, but the reconfiguration that this performs points back to the body as its source. In this sense, then, a technical element "has always inhabited and mediated our embodied coupling with the world."[90] Whereas Butler—as discussed in chapter 4—understands this coupling to be the continual and reiterative process of subject formation, though, Hansen understands it as the exteriorization of an anterior subjectivity. That is, because Hansen takes the body to be "invariant," he is able to mobilize a topological analysis of it. In so doing, he figures a gap that is fundamentally different from the deconstructive gap: where the former is a spatial separation, the latter is both a performative gap and a disidentification (hence *différance*'s etymology in both "difference" and "deferment").

What is notable in this configuration is the conflation of technical and embodied reality that it enacts. Simply put, the distinction between virtual reality and so-called flesh reality is effaced, for Hansen, so that "all reality is 'mixed reality.'"[91] In short, "there can be no difference in kind demarcating virtual reality (in its narrow, technicist sense) from the rest of experience"[92] because all experience pertains first to an embodied subject. As such, virtual reality is not so much something that is produced by technology as it is a "biologically grounded adaptation to the newly acquired technological extensions provided by new media."[93] Virtuality, then, is not only interpreted by analog bodies but is actually included in the constitution of these bodies as such.

In this context, Hansen insists on the cultural significance of virtual reality as "our culture's privileged pathway for . . . exposing the technical element that lies at the heart of embodiment":[94] paradoxically, the value of virtual reality (again, in its narrow sense) is not found in the purported new freedoms that it offers but rather in the way that its physical latitudes lay bare "the enabling constraints of the body";[95] virtual reality articulates the body's necessity. Here again, it is precisely because technics is included within a primordial notion of embodiment that we can comprehend virtual reality as a reality at all, which is to say, as having enactive capacities rather than simply representational or simulational ones. In a sense, then, Hansen is simply staging a performance of deconstruction in and as embodiment.

In this context, Hansen argues in *Bodies in Code* that mixed reality is a transcendental condition for human experience. That is, mixed reality designates an "'originary' coupling of the human and the technical" that grounds experience as such and that "can only be known through its effects";[96] whereas new technologies, specific historical realities, and evolved biologies all play a part in determining modes and meanings of perception that are absolutely unique to a given individual, the fact of human perception itself—or human "information processing," as Hansen sometimes terms it—nonetheless always grounds this specificity. Put simply, though information processing may radically transform the instance of embodiment to which it points, the fact that it must be directed at an embodied reality always remains. Moreover, Hansen's conceptualization of "mixed reality" includes the assertion that this works both ways so that an instance of embodiment—a human, for example—necessarily operates (in the broadest possible sense) relative to its surroundings.[97] This omnipresent bivalent coupling, then, is precisely what Hansen designates with the term *mixed reality,* effecting a passage "from the axiom that all *virtual* reality is mixed reality to the more general axiom that *all* reality is mixed reality."[98]

This persistence of an embodied relation to technics (which is to say, of embodiment) is the sense in which mixed reality is both transcendental and omnipresent for Hansen. Moreover, this status means that the effect of technological innovation is not to create new realities but rather to alter the relationship between our perceptual apparatuses and the techno-embodied reality that grounds them. In this, Hansen marks a crucial departure from both the scientific–evolutionary technological posthumanism of Dyens and from the humanist technological posthumanism of Hayles. That is, in contrast to the former, Hansen does not understand alternate strata of reality—genetic reproduction, macrobiological swarming, microcomputer processing, and so on—to possess an otherness that is fundamentally inaccessible to us but rather takes them as emblematic of an otherness that is constitutive of human embodiment: incompatible realities are not incompatible with our conventional perceptions because they are inaccessibly distant (for Hansen) but rather because they are too close, they are the ground of perception. In this sense, contemporary technologies create different points of contact—quantitatively different in scope but to such an extent that a qualitative difference emerges—between humans and nonhumans that thereby shift the ground of perception to give the effect of new strata of reality. In addition, then, Hansen also does not accept Hayles's suggestion that fundamentally new realities emerge from the intermediations of humans and machines: though, epistemologically, Hansen characterizes the historical evolution of humans technogenetically (as Hayles does), he nonetheless insists on an ahistorical mechanism underlying this process that grounds epistemological accounts of subject positions in an ontology of primary subjectivity. Thus, though both technology and embodiment are—to an extent—taken to be extradiscursive by both Hansen and Hayles, for Hansen, this extradiscursivity is specifically *pre*-discursive. Relative to both Dyens and Hayles, then, Hansen's position performs the crucial shift of acknowledging the particular worldview that undergirds the entrance of humans and technology into a joint discourse.

In this respect, our "age of total technical mediation"[99] is, for Hansen, the age in which the mixedness of reality is exposed. That is, reality has always been a mixture of virtuality and actuality, but contemporary technologies dramatize this fact in a way that has not previously occurred. In Hansen's (beautifully tenebrous) prose, the present historicotechnical moment is thus dubbed the "becoming-empirical . . . of mixed reality as the transcendental-empirical" such that contemporary technologies empirically manifest "the condition for the empirical as such."[100] Paradoxically, then, Hansen insists that the disembodying tendencies of digital technologies serve to make perceptible the embodied reality—which includes technicity—on which they depend.[101] As such, the shift to mixed reality effected in and by contemporary technoculture is a paradigmatic—rather than ontological—one.

If Hansen insists that mixed reality transcends local technological conditions, he does not by this intend to diminish the importance of the latter. Indeed, by making mixed reality perceptible, and widely so, digital technologies bring "an opportunity to revalue the meaning and role accorded the body within the accepted conceptual frameworks of our philosophical tradition."[102] Thus Hansen notes that today's exemplary mixed-reality situations—"interrupting a meeting to get data from a digital database, comparing a two-dimensional architectural drawing with a real-time three-dimensional visualization, acquiring an image of oneself through the social prosthesis of common sense that is contemporary television"[103]—are not intended to produce an "immersive" experience, in the sense of simulating reality. Instead—as we saw in his reading of *skulls*—Hansen argues that these situations each utilize the "capacity of our embodied form of life to create reality through motor activity";[104] the important thing is that, prior to the exposure of mixed reality through digital technologies, this capacity was entirely unremarkable. Thus, by moving perceptual experience beyond a single experiential frame, the question of "what makes . . . passage from one realm to another so seamless"[105] can be asked.

In the context of the discussion being presented here, the point is that Hansen's understanding of embodiment and technicity (as being constitutively linked) necessarily reconceptualizes embodiment in a way that avows their connection, while still retaining embodiment's primacy as the ground on which this conflation takes place. I have already argued that this understanding is more complicit with deconstruction than Hansen suggests, but it bears noting that what is thus at stake is not only a prelinguistic body but also an articulation of the "spacing" that comprises an organism's embodiment and ultimately provides a framework for shifting from a theory of affection to one of affectivity. In this, Hansen promotes an understanding of the body that is based in its operational perspective.

This rendering of embodiment is implicitly underwritten by an emerging (and still contested) cognitive theory of conceptual integration called *double-scope blending*. In essence, this theory argues that humans are the only species capable of thinking and feeling beyond the scale of their biological configuration and that this is accomplished through the ability to use our human scale as a scaffold ("a cognitively congenial basis") from which to reach out, manage, manipulate, transform, develop, and handle vast conceptual networks. Importantly, this "network scale" can be vast, even though the human scale is not, because it is anchored in a human scale. In turn, "these new human scale blends become second nature for us, and blending is recursive: packed, human scale blends become inputs to new networks [so that] what was once beyond human scale is now packed to human scale."[106] In this way, double-scope blending supports Hansen's operational approach by suggesting that "basic structures already present in our sensory-motor processing can be recruited for abstract thought without presupposing separate systems allegedly unrelated to our bodily engagement with our environment."[107] In short, this view holds that we are organismically configured to produce representations, but as a result, these representations remain grounded in our operations as a biological organism.[108]

Hansen explicitly draws the logic of this operational perspective from contemporary autopoietic theories of cognition. In particular, the "enactive" approach promulgated by Varela is apt, as it is based on situated, embodied agents. As Varela explains, this approach to cognition "comprises two complementary aspects: (1) the ongoing coupling of the cognitive

agent, a permanent coping that is fundamentally mediated by sensorimotor activities; and (2) the autonomous activities of the agent whose identity
is based on emerging, endogenous configurations (or self-organizing patterns) of neuronal activity."[109] For Hansen, then, considering embodiment
in this way offers the opportunity to consider a body as a bounded entity
that is coupled with technics, without falling into the reductive binary
logic of (non)identity: the body is coupled with technics in precisely the
sense that all life "necessarily involves a 'structural coupling' of an organism and an environment."[110] In this approach, it is movement, rather than
identity, that "takes center stage as the act which, in any specific context,
correlates, articulates, or mediates between space and time."[111] From this,
Hansen turns to operationality to provide a means of articulating the body
as an "originary condition of real experience."[112]

And yet (once again), Hansen's insistence on the term *origin* bears
further reflection: what is the body's "originary technicity" if not precisely
an example of "origin-heterogeneous," as described by Derrida?[113] Indeed,
in showing that the body's movement "speaks to a modality of life that
lies between and conjoins—that composes—space and time,"[114] Hansen
would seem to be reinforcing the Derridean truism that every identity—in
fact, every term—is necessarily infected by others that it cannot register:
technics itself, for Hansen, is placed "neither on the side of consciousness,
nor on the side of matter, but rather as their mediation, the transduction
to which they owe both their coupling and their proper existence."[115] I
raise this concern again because Hansen's response to this critique shifts
in *Bodies in Code*: whereas *Embodying Technesis* focuses on specifying what,
precisely, technology is, *Bodies in Code*'s operational emphasis attempts
instead to specify *how* embodiment—as the coupling of a body and technicity—produces the effects it does. That is, Hansen is less interested at
this point in what humans and their embodiments are (as categories) than
he is in what is entailed by these constructions. In this light, claiming an
originary status for a (subjective) body amounts to naming it as an enabling
constraint for the enactive approach that Hansen undertakes. Whereas a
deconstructive sensibility prevents us from agreeing with the definition of
a presubjective body that Hansen mounts in *Embodying Technesis,* Hansen's
shift in emphasis might permit him to sidestep these concerns because he
is no longer delineating a fully present body per se.

Indeed, Cary Wolfe convincingly argues in *What Is Posthumanism?* for a view of systems theory (specifically that of Luhmann) as the "reconstruction of deconstruction,"[116] noting that Derridean deconstruction and Luhmann's systems theory each approaches a similar problem from opposite directions. Specifically, he cites Schwanitz to note that both approaches

> make difference their basic category, both temporalize difference
> and reconstruct meaning as a temporally organized context of
> displacement and deferment. Both regard their fundamental operation
> (i.e., writing or communication, respectively) as an independent
> process that constitutes the subject rather than lets itself be
> constituted by it.[117]

This is not to efface the difference between the two approaches—Wolfe argues that systems theory links deconstructive dynamics "to their biological, social, and historical conditions of emergence and transformation"[118]—but only to point out that Derrida is not necessarily opposed to the operational logic that Hansen espouses.

However, this line of argumentation does not ultimately obtain in Hansen's case because he continually extends the import of his thought beyond this constraint, which is to say, beyond a deconstructive reading that would understand embodiment's operational "originality" to also indicate its nonoriginality.[119] Specifically, Hansen's shift toward an analysis of effects depends on the possibility of extralinguistic meaning from which it moves away so that his disavowal of grammatology might be considered a "meta" enabling constraint that allows constraint to mean something. In this context, then, one wonders if the term *origin* doesn't thus function as a rhetorical bivalence in the text, at once indicating Hansen's insistence on following through on a specific, organismically organized account of technological posthumanism, while simultaneously acknowledging—rhetorically—the limitations of doing so. Whether this is a conscious tactic on Hansen's part is, perhaps, irrelevant: instead, the ambivalent rhetoric of the text points to an opening in it that suspends, if not extinguishes, a deconstructive line of critique.

Indeed, the operational emphasis that becomes increasingly prominent in Hansen's work moves his thought ever closer to the linguistic ambivalence with which he continues to hold it in contradistinction. Perhaps most telling, in this respect, is his conception of "bodies in code" as a designation of the way that embodiment "is necessarily distributed beyond the skin in the context of contemporary technics."[120] Thus a body in code is not a computational body in the sense that Hayles's position critiques, nor is it a simulational body in the sense that Dyens's scientific technological posthumanism (ambivalently) performs. Instead, a body in code is "a body submitted to *and constituted by* an unavoidable and empowering technical deterritorialization."[121] As with mixed reality, then, all bodies are "bodies in code" because coding indicates the (technical) process through which bodies are coupled with their environment.

Here again, though, the ambivalence of this deterritorialization is crucial because its "technical" aspect indicates the constraining role that technology plays in addition to its more regularly cited extending capacities. Indeed, as Hayles notes, this insistence on the constraints that remain active in digital bodies is at the heart of one of Hansen's significant points of divergence from Deleuze, his frequent interlocutor. Summarizing Hansen's argument, Hayles notes that Deleuze and Guattari serve their own philosophical commitments at the expense of reality, doing away with all constraints whatsoever by refusing to recognize the constraints built into self-organization when it takes place in a biological domain.[122] In contrast, Hansen suggests bodies in code as a means to understand these constraints not only as enabling individual embodiments but also as part and parcel of the complexity—in the full, evolutionary sense of the term—from which emerges the new organizations and distributions of the senses that suggest technology's importance in the first place.

Hansen thus argues that through coded interactions with the environment a body takes on an agency that is at once constrained and distributed. To elucidate how this is the case, he refers to Merleau-Ponty's distinction between the "body schema" and the "body image," each of which offers a mutually exclusive way of conceptualizing the body: whereas the latter pertains to an apprehension of the body as an external object, the former privileges the (autopoietic) operational perspective that Hansen suggests is paramount.[123] Importantly, Hansen emphasizes that "body image" is somewhat of a misnomer, as his argumentation (if not his argument per se) hinges on its referring to something "much richer . . . than a mere

representation,"[124] acting instead as the object of intentional consciousness such that—put simply—to the extent that we apprehend ourselves,
we do so through a body image.[125] In contrast, the body schema refers
to that which grounds experience as such, in the sense that it refers to
the particular configuration of the body through which phenomena are
made sensible. Put differently, the body image emanates from the body,
extending it, whereas the body schema is "a kind of infraempirical or
sensible-transcendental basis for intentional operation."[126] Though both
refer to the body, then, they do so from "opposite sides of the intentional
relation [such that the body schema involves] an extra-intentional operation carried out prior to or outside of intentional awareness."[127]

And yet, though the body image and body schema do not overlap,
they are also not of equal ontological importance for Hansen. That is, the
operational perspective of the body schema means that, by definition, it
relates to its environment: here again, Hansen's autopoietic conformation
dictates that organisms react to environmental perturbations by modifying
themselves in a way that preserves the distinction between their identity
and their environment, in the sense that an autopoietic system is one that
"continuously produces the components that specify it, while at the same
time realizing it [i.e., the system] as a concrete unity in space and time
[that] makes the network of production of components possible."[128] As
an autopoietic system, then, the body schema includes the technics of the
body image as the process through which it is exteriorized such that, in
an important sense, the body image is a derivative emanation from the
body schema. Thus Hansen's understanding aligns with Varela's observation that reproduction "must be considered as an added complexification
superimposed on a more basic identity, that of an autopoietic unity, . . .
[because] only when there is an identity can a unit reproduce."[129] The
point for Hansen, again, is that the separation of the body from itself that
takes place as the body image requires a preexisting body that is identifiable in its own right. Here again, then, Hansen's departure from Hayles is
marked: for Hayles, the body schema would intersect with the body image
according to her revised version of the semiotic square[130] and thus would
remain within the realm of signification as such; for Hansen, by contrast,
the body image feeds back to construct novel points and means of access
to the body schema but does not challenge the latter's ontological priority.

IF I HAVE SHOWN that Hansen only tells half the story, his argument for an organismic understanding of technological posthumanism nonetheless remains an important specification—still within the play of linguistic representation—of embodiment. In this, the critical thing to note is how Hansen has shifted the mark. It is true that he—to a certain extent—has successfully articulated embodiment from an (organismically posthumanist) operational perspective, but it is not clear how the bounds of the human are delineated, except from an anthropocentric perspective that is constructed linguistically. That is, Hansen succeeds in accounting for technology's relation to the human sensorium, but in so doing, he sets aside the question of how this sensorium accrues an economy of meaning beyond simply maintaining its operational integrity.

With this in mind, I will conclude this chapter by considering the most explicitly political moment in Hansen's oeuvre to date, chapter 3 of *Bodies in Code* (titled "Digitizing the Racialized Body, or the Politics of Common Impropriety"), in which Hansen argues that all humans are bound by a common impropriety that can be understood as a form of resistance to the reductive classification of people relative to categories of social visibility. In particular, Hansen's analysis of "Internet passing"[131] is illustrative of the pure body-in-code: because online self-invention yields an identity that is "an imitation of an imitation [i.e., an imitation of culturally sanctioned signifiers]," he insists, these identities are "purely disembodied simulacr[a]."[132] Precisely because of this status, though, Hansen argues that, for example, in the case of racialization, "a certain *indetermination* in the correlation between racialization and the image [is introduced, suspending] the 'overdetermination' of the black body 'from without' and thereby position[ing] the image as a static fixation of individuation."[133] In short, then, Internet passing catalyzes an experience of *failed* interpellation—of an incongruity with oneself—that is definitive of affectivity (i.e., of primary embodiment as the grounding space, or "spacing," of perception). Thus, as Jennifer Gonzalez notes, Hansen presents a pedagogical vision of cyberspace in which the transcendence of visibility teaches those who are engaged in online passing the "bankruptcy of categories of identity."[134]

The properly political dimension of Hansen's reading of digitality, then, obtains in the fact that "online interpellation submits everyone—not just a particular subgroup—to the condition of having to pass; the generalization of the *condition* means in effect that the resulting abjection of the lived body from the space of intelligibility (visibility) can no longer be limited to certain subjects, but rather becomes a problem for all."[135] And this, Hansen argues, points to the shortcomings of existing theories of interpellation: in Butler, for example, performative practices "function by making the bodily residue [or representation] culturally intelligible"[136] and thereby build the radicality of embodiment into the scene of interpellation and thus deny it any standing outside the interpellative act. By contrast, the shared experience of Internet passing means for Hansen that

> *we all must live the erasure of our lived bodies,* [so that] what is most significant about the transcendence of visibility in online interpellation is less the possibility it affords for new modes of *represented* agency than its exposure of the violence exerted on bodily life by generic categories of social intelligibility and by the politics of recognition— identity politics—that it subtends.[137]

In short, then, Hansen argues that Internet passing does not so much open new avenues for self-expression as it rearticulates the bodily constraints that condition such expression in the first place.

Clearly Hansen's analysis of Internet passing restages his argument against "linguistic reduction" in a political dimension. Here again, then, it merits note that the meaning that Hansen ascribes to this politics remains contingent on the (representational) process of meaning-formation that he seeks to subjugate. However, we (like Hansen) can further specify this objection in its political aspect by considering his assertion that digitality makes intelligibility "a problem for all." Bypassing the obvious objection that would insist that, in fact, access to the Internet is not uniform or ubiquitous but is instead subject to its own complex social economy,[138] we might ask what the term *problem* means in this instance. That is, why does the body require this linguistic intelligibility such that it constitutes a problematic in the first place?

In fact, given Hansen's larger project of freeing embodiment from linguistic interpretation, it would seem that just the opposite would be desirable in a bodily economy: to the extent that Hansen argues that bodily specificity (including its technical element) is violently reduced by technesis, it is not at all clear why this entrance into discourse—the condition of intelligibility—is politically desirable. That is, what mechanism (other than language) allows bodily concerns to migrate from the level of a technobiological organism to that of a political organism *without abstracting the contents of these strata into the logic of the body?* In short, if an operational perspective constructs an organism in relation to its environment, how is the material specificity of that environment registered in its own right (i.e., as an extralinguistic radical exteriority), while simultaneously being related to its constitutive organisms in their own right? That is, how do we register the material specificity of performativity, since it is clear that "not all forms of performance are equal, nor do they have equal effects."[139] In the context of grammatology, this paradox is recognized to be constitutively irresolvable and is acknowledged through the (hyper) mediating role afforded *arche-writing*. If Hansen shifts the site of this mediating role to the practices of bodies, though, it is not clear how intersubjective and interstrata political agencies can take hold without abstracting bodies into a discourse that (according to Hansen) does not register their specificity.

Further to this, then, we can conclude that the embodied excess to social intelligibility that Hansen explicates robs that intelligibility of that which makes it, literally, desirable. That is, if the conventional engine of interpellation is desire (which is predicated on a fundamental and radical lack that is articulated in and as the linguistic dimension), then the politics of Hansen's organismic technological posthumanism comes to its crux: on one hand, Hansen cannot follow Deleuze and Guattari (for example) in refiguring desire as a line of flight that attaches to real objects because this would preclude the technobiological constraints that ground affectivity and that motivated his intervention in the first place; on the other hand, Hansen cannot maintain these constraints without invoking a notion of desire predicated on a lack that is linguistically inscribed by the Law. In the absence of either perspective, then, it remains unclear how political agency can come to be registered as a problem in the first place.

Thus the politics of Hansen's organismic posthumanism do not spring from the primordial subjectivity that he claims but rather from the claiming itself: in the case of political agency, Hansen's writing performs the wrangling that has come to characterize the relation between the discourses of desire alluded to earlier. Indeed, this orients us toward a broader observation about Hansen's understanding of affectivity, one that points back to the problem with which this chapter (and Hansen's theoretical oeuvre) started: in figuring a prediscursive body, Hansen in fact figures the discursive rent between technologists and cultural theorists that he seeks to address. Thus, because he fails to take the discourse of technology beyond the pale of linguistic analysis, Hansen succeeds in articulating the necessity of continuing to think deconstruction in its most radical dimensions. That is, the rifts opened by Hansen's insistence on the primacy of affectivity remind us that the categories that language produces (such as race and gender, in the case of Internet passing) cannot simply be "moved beyond" because they are not properties of individuals but rather "relations of public encounter."[140] In this, a paradox takes hold: precisely because theoretical concerns do, necessarily, subjugate practical ones with respect to generating a meaningful understanding of technological posthumanism, they themselves are also subjugated by a profound co-implication of practice and research that precludes in advance the possibility of either practice or theory keeping what it wins from the other. This paradox is that of deconstruction and is precisely what is foreclosed by the misreading of Derrida that allows Hansen to clear the ground he then claims. By showing the ambivalence that grounds Hansen's thought, then, I hope to reveal the "leaps of faith" into sensibility that take flight to redouble linguistic contingency under the auspices of eliminating it.

IN ADDITION to what I have offered previously, the challenge that deconstruction poses today is also one of comportment. That is, if I have taken some pains to insist that Hansen hasn't simply misread Derrida but actually *performs* the ambivalence that deconstruction diagnoses, this is partially to buttress my argument against the notion that I am trying to execute a conservative gesture of "taking Hansen back to Derrida." To think deconstruction in its full radicality would involve, equally, a sense of deconstruction moving forward through Hansen; the renewed urgency that Hansen lends the discourse is, after all, no small matter when one acknowledges that Derridean deconstruction—like any discourse—morphs, coalesces, accelerates, and slows according to forces in addition to those of reason.

6 SKEWED REMOTE MUSICAL PERFORMANCE: SOUNDING DECONSTRUCTION

> The point is that what transcends that reduction and schematization is not a substance, content, presence, or place ... but rather a "beyond" ... that is at the same time radically intimate, a beyond that is not, in Derrida's terms, a place. In short, the *transcendent* must be rethought as the *virtual*.
>
> CARY WOLFE, *WHAT IS POSTHUMANISM?*

If the organismic technological posthumanism discussed in chapter 5 performs an intensification of the paradoxical (deconstructive) causality that it disavows, this chapter discusses an art practice—Skewed Remote Musical Performance (SRMP)—that makes this performance explicit. Moreover, this practice connects the ambivalence of deconstruction to a digital network without using visual representation. In so doing, SRMP performs a deconstruction of the organismically configured body that exists, for Hansen, abstractly and prior to sensation as such: in place of primary embodiment, then, this chapter (via SRMP) offers sonic relation. To this end, the chapter reaches beyond its grasp to articulate the sound of SRMP as a "field-mosaic" in McLuhan's sense, indicating a paradoxical relationality as the "origin" of the terms related. In particular, the chapter's analysis of SRMP articulates a notion of sound that reaches toward the "fieldness" of this relationality precisely by refusing to give sonic instantiations ("sounds") primacy. In this refusal, the "fieldness" of SRMP's sound also prevents us from registering a human organism prior to its relational status: in the digital echo chambers of SRMP, embodied organisms do not dominate their representations but rather coexist with them via complex intermediating networks. Ultimately, then, SRMP models a way in which sound disjunctively intervenes in constructions of presence and absence, opening the body to a relational play that not only moves between those two poles but also constructs them as poles even as it is constructed by them.

Accomplished through the open source software SuperCollider, SRMP consists of two remotely situated musical performers who collaborate in real time via a computer network.[1] The defining characteristic of SRMP is a skewing mechanism that results in a situation in which the sounds heard in each of the two locations are markedly different from one another and specifically differ such that the details of their differences are not anticipatable or captured. For example, in the initial performance—which took place simultaneously in San Diego, California, and Victoria, British Columbia, Canada—each SRMP performer could articulate any of an array of sound files using a musical keyboard interface that allowed him to select the sample, select an effect chosen from a bank of signal processors, select the parameters of the effect, and indicate the articulation's duration and rhythm.[2]

To introduce a significant difference between what was heard at the two locations, the computer randomly turned on and off a skewing mechanism that, in this case, altered which sound file was played at the remote location but did not alter the effect applied to the sample locally (see the diagram of SRMP's signal processing). Theoretically, the skewing that is used in SRMP could be applied to any sonic parameter (or visual parameter, for that matter) that the software can recognize (which is to say, any parameter that can be digitally manipulated). In practice, the skewing used in SRMP has tended to combine a series of binary choices (i.e., whether to apply a sound effect at all; whether to substitute a randomly chosen effect for that which was remotely specified; whether to apply the effect uniformly or only to certain sound samples; whether to articulate the effected in addition to the original sample or as a substitute for it; etc.) with a continuum of dynamic specifications (i.e., how small will the "grains" in the granular processing be; to what extent is pitch shifting applied; what are the relative volumes of the processed samples to the originals; etc). This is to say that the skewing does not introduce new parameters to the performance but rather alters the sound of the parameters already included in the performance interface. Similarly, skewing has not been applied to formal compositional structures of relation in these performances but functions instead at the level of the performance itself. With

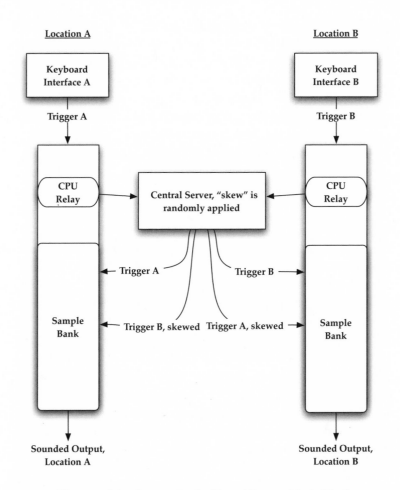

FIGURE 7. Diagram of signal processing for Skewed Remote Musical Performance.

all this said, the most important things to note are that skewing happens between both locations, the sound is never skewed locally, and the local performer does not know if or in what specific way the sound is skewed remotely.[3] To further emphasize this skewing, SRMP is typically presented as a structured improvisation, with the specific structural parameters varied from performance to performance.[4] (In this sense, SRMP indicates a performance medium rather than a specific piece.[5])

One of the motivating factors in SRMP's development was a perceived need to address the problems that recordings raise for live (sonic) performance, particularly in the case of electronic instruments. That is, because recordings offer the opportunity to redo, edit, and select sound material, what does a live performance offer the audience that it cannot get from a recording, particularly since audio recording and playback technology have progressed to the point where recordings cannot necessarily be acoustically differentiated from their live counterparts?[6] In the case of acoustic instruments, one answer to this challenge is that the specificities of a performance unfold relative to the "energy" of a given concert setting, a formulation that sounds vague until we remember that the physical acoustics of a space—including the changes to the acoustics that result from audience size and configuration—directly relate to the sense of timing that an instrumentalist develops. That is, a sensitive instrumentalist will intuitively adjust her playing to the feedback that she receives from both the space and the instrument she is playing, in the same way that a sensitive conversationalist will adjust her speech to a given setting or interlocutor.

However, in the case of music that is articulated through loudspeakers (and particularly music involving the use of prerecorded sound samples), this responsiveness is far less available. Specifically, predefined sounds (be they synthesized or prerecorded) include a simulated (predefined) acoustic space that is inseparable from the sound and is thus piped into the live setting and cannot be adjusted on the fly to the nuances of the space.[7] As a result, musicians working with electronics often express the need for other ways of addressing the challenge of recordings, be they through shifts in theoretical focus (such as an emphasis on the social element of live performance) or shifts in concert paradigms that range from sound installation to alternate placements of loudspeakers throughout a hall to highly specified compositional approaches.

The point, then, is that the solutions that have regularly been offered as justification for live performance have tended to address this problem by shifting its content. That is, there are certainly reasons why live performance remains valuable, but these explanations do not address the underlying question of why it is valuable for live performance to be valuable in the first place. That is, why insist? Put differently, there have been numerous demonstrations of what acoustic venues have to offer over home audio but very few of what is acoustically specific to liveness itself for electronic music. If acoustic performance practice addresses this via the relation between a performer, her instrument, and the performance space, then, SRMP focuses instead on the inter- and intraperformer relations as they unfold in real time.

Ultimately, though, SRMP fails the task that it set for itself: there is no convincing reason why, from the audience's perspective, the performers could not just mime playing their instruments while a recorded track is sounded. (Indeed, we could even say that such "miming" is a significant part of conventional performance practice because live electronic music often features prerecorded samples in some form.) However, what emerged from this goal was a realization that SRMP does offer its performers a specificity of relation that is not typically (explicitly) available, namely, a materially specified experience of disjunction (rather than an abstractly disjunctive sensation). Thus what is notable about the use of a skewing mechanism in SRMP is not only the disjunction that it introduces between the two performers but also the discrepancy that it creates between a performer's action and the action's sonic representation in the remote location. In each case, a different mode of deconstructive gap is figured: in the first case, the skewing might be said to deconstruct the objective presence of the piece, in the sense that it results in a literal difference between the "piece" that is produced at each location (despite their both being created through the same performance activity). As a result, the minimal difference that always exists in signifying processes—the difference of spacing and timing that Derrida terms *différance*—is amplified and (consequently) avowed in the performance; in the second case, the skewing makes explicit the nonidentity of each performer with his actions because the performers' musical activities result in multiple and unpredictable results. In this, signs of the performers' presence are accompanied by marks of their absence so that the actions are not theirs per se but rather represent a certain degree of play. In both cases, then, the work signifies a present absence that explicitly renders the piece materially contingent.

In each of these cases, SRMP performs a different articulation of the deconstructive gap, aligning with two prominent ways in which this gap has regularly been figured. What bears emphasis here, though, is the fact that these absences *are performed* in SRMP, which is to say that they are made present (though incompletely) through continual and reiterative significations of their absence. In this sense, what is perhaps most engaging about SRMP is the material specificity that it lends this present absence: the constitutive paradox of deconstruction does not function as an abstract category in SRMP but is instead rendered in a way that specifically and dynamically relates to the host of other factors that are active in a given moment. In short, SRMP is a profoundly practical rendering of the ambiguity that deconstruction so aptly describes.

Indeed, this trait is evidenced in the performers' ability to develop a performance syntax with one another, despite the fact that they do not know whether (or in what ways) the remote signal is skewed.[8] Thus, for example, one performance featured a precomposed formal transition that was to be indicated by the remote performer's execution of a large expansion in the "grain size" of a sample: if the remote performer executed the gesture over a period of two to five seconds, the local performer would understand this as a cue to shift to a new, predetermined, texture; if, instead, the expansion was made over a period of nine to eleven seconds, the cue would be to a different predetermined texture. The important point, then, is that recognition of the gesture was crucial to a successful performance in this instance because the composition required this recognition to emerge as a composition: in an otherwise improvised context like the one in this example, these types of composed transitions are key formal constraints. In this case, this recognition was successfully achieved when the local performer heard a decrease in volume that was executed over one second and repeated three times because he was somehow able to recognize this as a skewed rendition of the established cue.

What is crucial in this example is that the performers' success did not stem from their recognizing and identifying the literal sounds but rather from their listening to the "presencing" of the two modes of absence described earlier. In this sense, then, the performers in SRMP interact by listening to sounds that are not sounded, specifically hearing them as alibis for soundings that do not sound. In fact, the performance could not have succeeded if the performers had relied on their ability to recognize the literal sounds because those sounds that were preordained with formal significance were never literally audible. What SRMP affords, then, is a rendering of the play of deconstruction's presence and absence that gives equal and simultaneous weight to each intensity, thereby continually re-vivifying the paradox that exists in and as the registration of meaning (in this case, musical meaning, in the sense of identifiable formal transitions).

What is performed in SRMP, then, is a particular way in which a digital technology amplifies an existing paradox such that it may be explicitly acted on. Thus SRMP unfolds the implications of this technology in two distinct registers. On one hand, the technology functions in the perfor-mance as an abstract formal category that contributes to the undermining of conventional worldviews predicated on locating "the subject of speech in the same ontological space as the speaking subject."[9] In this, Hansen's critique is incisive and cogent and offers a model by which to understand both the importance and the limits of this critical stratum. However, what SRMP also robustly specifies—and what Hansen's perspective neglects—is the network character of its particular technology by maintaining the deconstructive paradox as it exists in the medium of sound. That is, by necessitating an awareness of sounds that are not sounded, SRMP's tech-nology makes explicit the gap between literal sounds and the sonic ontol-ogy in which they exist, while simultaneously pointing to the inadequacy of an identity-based notion of presence–absence in describing what is at stake in the practice. In N. Katherine Hayles's terminology (discussed in chapter 3), this amounts to an insistence on supplementing the continuum of presence and absence with that of pattern and randomness.[10]

By offering the means to explicitly perform this paradox, then, SRMP aligns with Hansen's argument that digital technologies present an op-portunity to expunge the technical gap between a subject and its exter-nalizations and to thereby effectively replace representation in favor of a "representative function." However, whereas for Hansen this difference moves the discourse from representing an external reality (i.e., represen-tation) to serving to reveal the primary interiority that is projected as representation (i.e., representative function), SRMP's decentering of the performing subjects suggests instead that this representative function is enacted via its play of representations rather than prior to it. Thus, though a grounding of representation in embodiment is indicated by the performers' experiences—which "do not take the form of a (representational) image but rather emerge through the *representative* function of the data"[11]—the inverse is also evinced by the dependence of the performers on represented identities to exercise their agency. Again, then, SRMP suggests that sound is anterior to its literal instantiation, but only in the sense that it indicates a relationality that emerges from the latter; in this paradox, a play of representation and representative function takes place.

To elucidate this point, certain aspects of McLuhan's analyses of the phonetic alphabet remain useful (despite Derrida's grammatology). Specific to the argument here, McLuhan discussed how the phonetic alphabet's division of language into vowels and consonants makes vowels a "percept without a concept" and consonants a "concept without a percept."[12] As such, the medium of the phonetic alphabet isolates the "nonsound" of a consonant in language, thus isolating a nonsound that is nonetheless heard. The point, for McLuhan, is that the formal structure of visual space—the space of presence–absence—always involves the interiorization (or suppression) of ground as a guarantee of abstract, static uniformity. That is, simultaneous to the introduction of the phonetic alphabet is the foreclosure from perception of the excess that grounds it: the introduction of an economy of language in which phonemes might be detached, manipulated, and exchanged without losing their identity sublimates the culturally specific meanings that inhere in linguistic practice.[13]

Literal sounds, as (visual) figurations, likewise suppress the relationality of sound so that we might think of SRMP's "sound that isn't sounded" as a concept without a percept. However, the crucial difference between the two media—that is, between language and sound—is that the simultaneity of SRMP's digital technology allows sound to be acted *even in the absence of its being perceived*. Indeed (and paradoxically), if the disruptive force of SRMP's skewing were less reliable at the level of perceptibility—if, for example, it introduced a perceivable lag when it was activated—the performers would be less likely to interact successfully: the disjunction that the skewing introduces into SRMP's literal sounds must be virtually perfect (i.e., must separate the literal sounds from their provenance in a way that completely hides the act of "cutting") to desublimate the relationality that grounds the performance (and is grounded by it). In this sense, the figurative separation that the skewing mechanism introduces into SRMP is precisely what allows the performers to act on the paradoxical sonic relation—which is both non- and hyperrepresentational—that binds them. Thus the skewing in SRMP mobilizes literal sounds as *representative functions* of sound rather than of an embodied subject so that the latter's relationality is heard—if not literally foregrounded—through them. As a result, the relationality of sound that SRMP deploys does not result in a perspective that denies the specificity of the performers' bodies (as Hansen might suggest) but rather unpacks the way in which the constraints of Hansen's organismic technological posthumanism overdetermine the sitedness of these subjects.

In emphasizing the performers' and computers' relationality, SRMP's deployment of its technology performs its specific networked materiality by deploying the difference between sound and literal sounds as an active one. In a sense, this difference is simply that between a sonic ontology and the literal sounds that populate it. Alternately, we can understand this difference through Steven Jones's argument that in contrast to "the traditional view that sounds signify an event (i.e., the sound *of* something, such as that of a door opening), recording technologies . . . shift this such that [a] sound now signifies 'the sound of something' (rather than signifying the sound of 'something')."[14] In a sense, then, literal sound as it is exchanged in discourse is equivalent to Jones's "something," whereas the discourse of sound coincides with "the sound of something." That is, literal sounds are identifiable, whereas sound is a continual process in which these identities are understood to exist in a mediated relation with one another.[15]

More broadly, the distinction between literal sounds and a sonic ontology can be expressed in the claim that literal sounds are figures (i.e., "sounds"), whereas sound is the specific mode of relation that grounds them. In this way, SRMP displays the paradoxical nature of sound's relationality, simultaneously insisting on its disjunction from literal sounds and its constitutive connection to them. In this sense, "sounds" foreclose their ground—that through which they are figured—to register themselves, but this characterization means that sound itself is not figured per se, except to the extent that it consists in the play of figures with a ground that is itself coextensive with its figurations. Put simply, sound is the medium of "sounds," even as each literal sound registers itself as an attempt to foreclose this medium by insisting on its own specificity. More than usual, then, the word *sound* is the death of the thing that it describes: via a process that might be called "auresis," literal sound freezes sound's indeterminate motions into "sound," or even "sound's indeterminate motions."

In SRMP, though, this reduction is also an expansion because it is only through its translation into discursively identifiable terms (i.e., into "sounds") that sound's relationality is made visceral, in the sense that it is imbued with meaning. Quite simply, the performers are able to hear—in the process of sound's abstraction into "sounds"—the movement of literal sound's absence into sound. Thus the crucial point is that sound is equally active in both sound and "sounds" so that the question of differentiating between the two becomes a question of registration rather than materiality. However, this distinction does not mean, as Hansen would have it, that aural space materially preexists its registration but rather that the differentiation itself is a symptom of discourse. That is, to say that a prediscursive materiality of sound cannot be distinguished from that of literal sounds is not to say that they are the same thing prior to discourse but rather that they exist precisely through their discursive correlation. As a medium, then, sound paradoxically preexists its content precisely because it also does not (and vice versa), which is to say that sound performs the irrationality of the real that Serres finds in all relation: the paradoxicality of sound and literal sound's relation is integral (i.e., the paradox would be redoubled if the relation were not paradoxical).[16]

It bears noting that this is in distinct contrast to Hansen, who takes a rational precedence of the medium—the fact that the body "exists as a unified field that [logically] precedes . . . the differentiation of the senses"[17]—to indicate the illogic of linguistic registration, namely, the fact that language depends on something preexistent that it must disavow (i.e., in Hansen's reading, the "primary tactility" of embodiment). As I argued in chapter 5, this claim is ironic because embodiment's precedence is itself linguistically figured by Hansen (both literally and in its use of visual logic) and thus depends on the very logic that it denies in order to signify. In contrast, SRMP puts the constitutive illogic of language—particularly its ambivalence with respect to identity—into play as a paradox. As a result, sound (as a medium) takes logical precedence over its contents to exactly the extent that it undermines this logic: if Hansen argues for the logical precedence of a body over its senses, then, SRMP counters by performing the illogical precedence of sound over "sounds." Put simply, SRMP does not contradict a perspective that would give embodiment primacy, except to show that such a perspective naturalizes the biases that underwrite it. In so doing, SRMP does not contrast Hansen's attempt to hold discourse accountable to embodiment by simply reversing the scenario but rather by first holding discourse accountable to itself.

This "illogic" points to the way that sound differs from Hansen's reading of tactility. To the extent that sound is distinguished from the visually figured "sounds" that it correlates, it closely resembles the notion of embodiment to which Hansen ascribes. Indeed, like Hansen's understanding of primary tactility (or "infratactility"), sound is here taken to be inclusive of the other senses.[18] Moreover, sound and infratactility both indicate a sensitivity that is external to the positive empirical figures of sense perception per se, and thus each indicates an unquantifiable yet specific materiality. However, by insisting on the irresolvable nature of sound's structuring paradox, an important difference obtains (despite their similarity in content) between the ways in which these terms operate. Specifically, I have chosen sound as my grounding metaphor in this chapter for two reasons: first, in the hope of precluding any confusion with respect to its ontological priority since, unlike infratactility, sound does not lay claim to prelinguistic meaning, and second, to take advantage of the unsi(gh)tedness that characterizes both acoustic space and our everyday experiences of "sounds"—as McLuhan learned from psychologist Carl Williams (who, in turn, worked with E. A. Bott), auditory space "has no center and no margins since we hear from all directions simultaneously."[19] Unlike prototypically tactile experiences, we are frequently disoriented by sounds. In sum, the metaphor—and both sound and infratactility are metaphors—relies on a shared experience of sounds as a moment of sensation that precedes identity, in the sense that everyday experience frequently includes both (1) hearing a sound before we know its origin and (2) not hearing a sound—consciously, as a sound—because it acts as a (seemingly) neutral carrier of a signal. In both these cases, then, sound differs from tactility in its character, if not its content.[20]

Indeed, these characteristics of sound relate directly to the medial specificity of SRMP. Imagine, for example, an equivalent setup using video rather than audio: such a performance would have to operate via entirely different aesthetic criteria because applying a skewing mechanism to a video would act to transform the image rather than acting on the gap between the performers' actions and the image. That is, real-time video tends either to project a subject directly (in, e.g., a video of someone or something), to represent its underlying agent via an avatar that is situated in a constrained world (as, e.g., in a video game), or to almost entirely sublimate its underlying agent (in, e.g., synthesized video of abstract, computer-generated images). In the first two instances, SRMP's skewing would not be effective because it would be perceptible so that it would deteriorate either the realism (in the first case) or the believability of the simulated world that accompanies this claim (in the second case). In the third case, the reverse problem would adhere: the skewing might be undetectable, but this would be the case because the relation of the video to its underlying agent was not established in the first place.[21] The point, then, is that sound tends to signify with a stronger ambivalence than other media with respect to the correlation of actors and their actions, in the sense that it simultaneously relies more heavily on signifiers that are detached from their alleged signifieds but also—paradoxically—takes these signifiers to be constitutively representative of something. As I noted in the introduction to this book, sound remains the conventional test of presence.

While sound offers an important specificity, though, its demarcation from other media is by no means either discrete or total; indeed, the parasitic nature of sound's semiotics—not to mention the constitutive relationality discussed earlier—means that sound is never just sound. While skewing video would not work in precisely the same way, for example, an analogous effect might be achieved through computer animation.[22] Ultimately, then, in the characterization of sound that I have offered here, I am not insisting on a hard and fast definition of sound. Rather, I am articulating a version of it that avows its necessarily metaphoric quality and thereby reminds the reader that the term is not intended to immediately indicate its object (which is not an object, in any case) but rather to instigate a language of the senses that is perhaps more suitable than conventional (visually biased) metaphors for characterizing the profound displacement that is experienced in SRMP.

In these senses, *sound* (as it is used here) is a more concertedly decentered metaphor than *tactility*, as it is understood by Hansen, and instigates a language of resonant relations in place of the latter's grounding hierarchy. In so doing, sound aligns with the network quality that characterizes digital technologies and is particularly suited to the claims to simultaneity that have dominated the latter's discourse. Indeed, McLuhan foresaw this connection between digitality and acoustic space in his frequent insistence on electric technology as a return of the acoustic, since the former obsolesces linear causality by acting at the speed of light.[23] Moreover, it is a peculiar irony of language that acoustic metaphors describe digital networks better than visual ones, because the hypercausality of digital networks (and electric technologies) operates via this condition of visibility. All told, then, this reading suggests that acoustic space might be understood as a reversal potential of vision, a moment in which its intensification reaches the point that it flips into its opposite.

Stated differently, SRMP performs an intensification of "sounds" that pushes a figuration of sound to its full intensity such that the virtual (or skewed) "sounds" cannot be distinguished from their real counterparts. In this, the performers themselves are to an extent virtualized as agents, not in the sense of their bodies being dematerialized (either idealistically or dystopianistically), but rather in the sense that a crucial component of their agency is sited in the machine through which they relate to one another. Indeed, in a kind of acoustic trompe l'oeil, one could even suggest that the performers' bodies act as the vanishing point of the machine's perspectival horizon such that they are rendered performance objects of a relational network that precedes them because it exists outside of the realm of precedence (i.e., as the speed of light).[24] In this, then, the performers are perhaps not so much virtualized as they are a testament to Deleuze's observation that "it is not so much that one cannot assign the terms 'actual' and 'virtual' to distinct objects, but rather that the two are indistinguishable."[25] As a result, to the extent that the performers of SRMP act in relation to the literal sounds produced, they do so within a virtual temporality consisting in a period of time "smaller than the smallest period of continuous time imaginable in one direction . . . [but] longer than the longest unit of continuous time imaginable in all directions."[26] This is the sense, then, in which SRMP constructs its performers less as agents than as a-causal resonances that perform the impossibility of a semiotics of sound per se that Shepherd and Wicke have identified: "sounds acting

as a medium [are] *materially* involved in calling forth from people elements of signification in a manner in which sounds as signifiers do not."[27]

The point, then, is that the digital technology of SRMP not only makes
the constitutive paradox of meaning described by deconstruction explicit
but also redoubles it in a language of technological performativity. After
all, though SRMP performs a formal undecidability, such a performance
nonetheless constitutes—*as a performance*—a kind of decision. In particular, this paradox is played out in SRMP as a radicalization of the claim that
sound is inclusive of vision, showing that McLuhan fails to follow the
full potential of his own thinking. That is, by arguing that "our private
senses are not closed systems, but are endlessly translated into each other
in that experience which we call consciousness,"[28] McLuhan suggests
(like Hansen) that there exists a closed system that precedes language
but nonetheless signifies.[29] By contrasting this claim, then, SRMP desublimates the (normative) leap of faith that is necessary to speak sensibly
of the "human." In SRMP, then, the human "is not now, and never was,
itself"[30] but instead operates precisely as the disjunction that maintains
this formulation.

Ultimately, then, SRMP relies on a primacy of sound over its literal
instantiations that can only be understood grammatologically. In this, the
primacy of sound is not an origin but the preclusion of origin through
which a primary relationality is retrieved. Put differently, SRMP emphasizes the deconstructive gap by performing it in an acoustic setting, thereby
sidestepping, if only for a moment, the visual dominance that has tended
to shape Western culture (including "Western art music"). That is, SRMP
makes deconstruction audible, and this audibility, the detritus of performance, ultimately represents the limit of sound's registration. Amid
this detritus, though, the subject that McLuhan posited as the source of
technological extension has multiplied and continues to multiply. What
SRMP performs, then, is not simply a deconstruction of "sound" as a positive term—or, for that matter, of embodiment as an originary and unilateral source of meaning—but rather the suggestion that artistic practices
might insist on themselves as conjunctive-disjunctions–disjunctive-con-
junctions: as sounds. When this is the case, the interval between subject
and object, the becoming of sound, is revealed as a paradox that invests
both terms.

CONCLUSION

REGISTRATION AS INTERVENTION: PERFORMATIVITY AND DOMINANT STRAINS OF TECHNOLOGICAL POSTHUMANISM

> In creation and observation works of art unfold as a sequence of events. But how? . . . [They] must be capable of generating both continuity and discontinuity, which is easier in reality than in theory.
>
> NIKLAS LUHMANN, AS CITED BY CARY WOLFE IN *WHAT IS POSTHUMANISM?*

> Perhaps these differences are superficial, perhaps they are destined to disappear. What is certain is that right now they do most obviously exist.
>
> SIMONE DE BEAUVOIR, *THE SECOND SEX*

A question that is often asked about technological posthumanism—and about virtuality, simulation, and a host of other developments that connect critical theory to digitality—is whether its subjects are something new or simply something that is newly understood. In this book, I have shown that there is no single answer to this question: Hansen, for example, argues that contemporary technologies offer new types of access to a (transcendental and) primordial subjectivity; Dyens, conversely, argues the reverse, maintaining that a fundamental shift takes place in the migration of evolutionary processes from genes to memes; and Hayles, for her part, occupies both positions, believing that "we have always been posthuman"[1] but also that new forms of life (and new subjectivities) emerge in contemporary intermediations of humans and machines. Clearly, then, the question of technological posthumanism's novelty is entangled with the particular construction of it that is being mobilized: presuming that a questioner already has a sense of what technological posthumanism is, for the questioner, any answer to this query would confirm or deny this construction as much as it would address the question per se.

For myself, this question is unanswerable for another reason, though, not only because of a perspectival bias that always inhabits binary decisions but also because it necessarily involves so many terms, which slide into one another. As a result, posing the question coherently requires precise and autonomous definitions of virtuality, technology, subjectivity, embodiment, life, reproduction, and the myriad other terms that obtain in this study. Simply put, such definitions are impossible to maintain, because doing so disregards the disciplinary contingency, inherent relationality, and simple creativity of their nominations; that is, doing so neglects terminology's status as "the poetic moment of thought."[2] More than nominating an undecidability of the human, then, technological posthumanism might foremost mark an (un)askability of the human, where the parenthesis indicates a necessity of asking that is put into play in and as the question's impossibility.

However, if the question of technological posthumanism's novelty cannot be coherently answered, the challenge that it implies—a demand to offer, or at least to gesture toward, something that cannot be neatly captured within existing analog terminology—has nonetheless been convincingly met. In fact, we might even say that the challenge is met precisely because the question remains unanswerable, in the sense that technological posthumanism has thus far refused the sedimentation of a unified discourse.[3] In this context, if a central task of a given strain of technological posthumanism has been (and continues to be) to register the changing relations of humans and technologies—or, more broadly, the complex interactions "between individuals as living beings and the historical element"[4]—this book has continually insisted that such registration is not (and cannot be) neutral. In particular, I have shown how three dominant strains of technological posthumanism each naturalize the process through which their articulation of the problematic takes place, thereby sublimating the ideology that they perform. Even the most skeptical critic—and skepticism toward the term abounds[5]—must agree that technological posthumanism, if nothing else, reinforces the extent to which humans are simultaneously object, subject, site, and actor of such performances. In this sense, technological posthumanism in general might be considered—following Hansen—as *humanesis*: a putting-into-discourse of the human. A crucial caveat in this formulation, though, is a reflexive quality that inheres in it: technological posthumanism is a putting-into-discourse that reveals that which is "put in" to have always already been discursively constituted.

Thus, although it is important to insist on the critical limits that obtain in the discourse of technological posthumanism, in the preceding chapters, I have additionally articulated what, in particular, is performed by setting these limits in the first place: for each of the three theorists considered, I have not only shown the trace of ambivalence that grounds his or her discourse but also how each theorist's "leaps of faith" into sensibility take flight to advocate ethically and politically—among other consequences, Dyens's denial of embodied reality produces a disavowed ontology of death; Hayles's emphasis on registering machinic code renders it subsidiary to (decidedly liberal) humanist values; and Hansen's willful misreading of grammatology produces a body whose linguistic contingency is redoubled. In each case, by emphasizing moments in which these strains of posthumanist theory turn back on themselves, I have articulated these logical gaps in their positive dimensions, showing their effects. In so doing, I have neither directly objected to the author nor posited an alternative position but have rather teased out the operations that make each version of posthumanism tick. Indeed, the intervening analyses of media artworks (in chapters 2, 4, and 6) are staged in this spirit and are thus intended to probe the particular theorist in question more concretely rather than to oppose him or her per se; that is, these chapters *listen* to the theoretical chapters, in the full (differential, parasitic, multiplicitous, and resonant) sense.

In this sense, I have interrogated the particular "technologies of self-hood" that Dyens, Hayles, and Hansen each perform, where each of the three holds in tension the enhancements, reversals, obsolescences, and retrievals of a particular configuration of humans and machines. And yet, if these three posthumanisms each mobilize a particular worldview (to use Hayles's reluctant term), they also interact with one another—both in and out of the confines of this book—to form a discourse of technological posthumanism. Thus, to the extent that this book coalesces this discursive activity into something meaningful, I myself am implicated in it because I have authored the boundaries of this discourse (as it is presented here). The point is that my description of these three technologies of posthumanism performs a fourth, a "discursive technological posthumanism" that can itself be considered through McLuhan's tetrad. Thus, via an analysis of Hayles, the values that human embodiment expresses are *enhanced* in human–technology relations, where all aspects of embodied human reality are indissociable from questions of technology but are not overdetermined by them. In this heightening, previous accounts of meaning-formation are lent new urgency and intensity. By contrast, we witness through Dyens the *obsolescence* of a scientific–evolutionary understanding of the posthuman such that the overdeterminations of the latter's empirical mechanisms are replaced by the type of intermediating feedback loops that Hayles so elegantly de- and in-scribes. Simultaneously, we are taken with Hansen into the deep time of technosubjectivity, where an iconic selfhood—preexisting the division of technics and embodiment—*retrieves* the urgency of deconstruction's scholarly intervention. And finally, from the coupling of my own narration of these scholars to sound-informed media art interventions, the *reversal potentials* of technological posthumanisms are desublimated as the politically charged assumptions that necessarily underwrite them. Taken together, then, the discursive technological posthumanism of this study is itself a technology of posthumanism, and each of its claims is necessarily contingent on the other three of McLuhan's fourfold intensities.

This fourth perspective (which is my own but is objectively inaccessible to me as a result) bears emphasis, if only because scholarly convention has precluded attending to it thus far in the book. That is, this book itself—or rather, the writing of this book—constitutes a Deleuzian assemblage of technological posthumanism, a rhizomatic multiplicity where (by definition) "any point of the rhizome can be connected to any other."[6] Because "determinations, magnitudes, and dimensions . . . cannot increase in number without the multiplicity changing in nature,"[7] it is understood that the conclusions of this project represent tracings that must be "put back on the map."[8] That is, the alliances traced here—be they similarities or points of diversion—occupy a plateau of the multiplicity "technological posthumanism," a "continuous, self-vibrating region of intensities whose development avoids any orientation towards a culmination point or external end."[9] Consequently, the subjects of this study are not presented toward a grand narrative of technological posthumanism that they either align with or not but are instead directly allied in the particular construction of technological posthumanism—the critical discourse of technological posthumanism—that is presented here.

What does this mean? It means that this book has used a narrow selection of thinkers to examine an indeterminate concept within provisional categories to elucidate a specific discourse that is brought into being through this process of description. What is offered in this book, then, fails to be Technological Posthumanism as it becomes a technological posthumanism, and nothing more. As Derrida has said, "it is between different things that one can think difference. But this difference-between may be understood in two ways: as another difference or as access to nondifference."[10] Though each of the theoretical chapters presented here has, in a sense, chosen the latter understanding, this book as a whole chooses the former.

Indeed, this multiplicitous nature is suggested by the artistic chapters presented in the study, which each take flight from the (seemingly) linear narrative of the book to suggest alternate readings that are not taken up further because they are beyond the latter's purview, though they certainly inform it. As these interventionist chapters suggest, then, there is simply no way around the contingency that haunts each construction of technological posthumanism so that it would be both futile and ill-conceived to attempt to definitively conclude this, or any, account of the discourse. Indeed, one of the predicates of this study is an understanding of language that avows its material effects, so such a conclusion would, by definition, undermine itself. Instead, this conclusion aims to reiterate the provisional status of the study, not in an apologetic sense of refusing to stand behind the claims that I have made, but rather as a simple statement of factual fiction—in the flavor of what Blanchot has called "fictive theory"[11]—intended to invigorate the granitic territory of a discourse described.

IT IS A TRUISM that art functions, at least in part, to create conditions in which audience members may "begin to question their own habitual perceptions and assumptions about being in the world."[12] This role is amplified in the case of media art because its very materials slide so nebulously between the worlds of aesthetics, capital, and politics. In this context, efforts to "explore how the medium/technology can deautonomize perception"[13] take on added importance: as we saw in Hayles's analysis of the Regime of Computation, naturalized technologies create metaphoric structures that dramatically condition thoughts, feelings, and actions and thereby directly affect our decision making. Ultimately, recognition of this contingency is a primary tenet of technological posthumanism in the sense that it marks the entrance of a human–technology coupling into discourse.

In the artistic practices discussed in this book, though, this role is taken up in a slightly different direction. Rather than attempting to access a prediscursive embodied Real, these pieces instead insist on the profound co-implication of discourse and embodiment by accepting it and even celebrating it. As a result, reversibility abounds so that we are continually reminded that language itself is captured within a paradoxical presence–absence that is, in a sense, embodiment. Thus it isn't that Hayles, for example, is wrong to insist on the impossibility of separating information from its embodiment but rather that this emphasis constructs—in the rhetoric of human values—the very separation that it opposes by acting as though disembodiment is a danger (and is thus possible). In contrast, this book insists that the question of embodiment is beyond the pale, not because it is outside of language (as Hansen argues), but rather because it is a paradoxical becoming of linguistic ambivalence.

Indeed, this active ambivalence is an especially necessary position to insist on today, as we move further into what IBM is calling "the decade of smart." I introduced this project by highlighting the ubiquity of technology and suggesting that we are in an era when understanding our relations to digital technologies in particular is a key component to addressing the question of agency in its contemporary appearance. This need is now all the more immanent, given that companies like IBM claim to have already answered it and are acting accordingly. Consider the following excerpt from a recent press release by the latter, promoting new "smart" technologies:

> Trillions of digital devices, connected through the Internet, [are] producing a vast ocean of data [that can be] turned into knowledge.

[Today, forward-thinking leaders are] finding the hidden treasures buried in data. Data is . . . revealing everything from large and systemic patterns . . . to the location, temperature, security, and condition of every item in a global supply chain. . . . That's a lot of data, but data itself isn't useful . . . unless you can extract value from it. And now we can.[14]

The point is that whether or not we are aware of it, we are implicated in this process of "value extraction" so that understanding this implication as a problematic—that is, as something that is simultaneously objective and subjective[15]—becomes paramount to any claims to agency we might make. Each of the artists and theorists of this study offers an idea of what this relation has been, is, and may become; one wager of this book is that understanding the connections, rents, and overlaps between these worldviews may shed light on the discursive movements they describe so that we may begin to swim through—and, perhaps, against—the dataverse with "smart" technologies of our own.

NOTES

INTRODUCTION

1   Brenda Laurel, as cited in Frances Dyson, *Sounding New Media: Immersion and Embodiment in the Arts and Culture* (Berkeley: University of California Press, 2009), 140.

2   Aden Evens, *Sound Ideas: Music, Machines, and Experience* (Minneapolis: University of Minnesota Press, 2005), 1.

3   Indeed, I think this difficulty in placing sounds is—in part—why we can still note the dearth of studies that focus on sound, even at a time when sound studies has both proliferated as an important element of disciplines such as film studies, media studies, and communications and had found traction as a discipline in its own right. I comment further on this distinction later in the introduction.

4   Cary Wolfe, *What Is Posthumanism?* (Minneapolis: University of Minnesota Press, 2010), 132.

5   Ibid., 196.

6   Eugene Thacker, *The Global Genome: Biotechnology, Politics, and Culture* (Cambridge, Mass.: MIT Press, 2005), 310.

7   Mark B. N. Hansen, *Bodies in Code: Interfaces with Digital Media* (New York: Routledge, 2006), 8.

8   Matthew G. Kirschenbaum, "'So the Colors Cover the Wires': Interface, Aesthetics, and Usability," in *A Companion to Digital Humanities,* ed. S. Schreibman, R. Siemens, and J. Unsworth (Oxford: Blackwell, 2004), http://www.digitalhumanities.org/companion/.

9   Lev Manovich, "Database as Symbolic Form," in *Database Aesthetics: Art in the Age of Information,* ed. Victoria Vesna (Minneapolis: University of Minnesota Press, 2007), 39.

10   In the context of this formulation, in *What Is an Apparatus? and Other Essays* (Stanford, Calif.: Stanford University Press, 2009), 14, Agamben takes the subject to be "that which results from the relation and, so to speak, from the relentless fight between living beings and apparatuses."

11   N. Katherine Hayles, *My Mother Was a Computer: Digital Subjects and Literary Texts* (Chicago: University of Chicago Press, 2005), 2.

12   Crucially, McLuhan (with his son Eric) indicates in *Laws of Media* (Toronto, Ont.: University of Toronto Press, 1988), 9, that these laws are "scientific"

in that they are testable, universally applicable, and yield repeatable results. Framed as questions, and intended to be asked simultaneously, the tetrad is: What does the artifact enhance, intensify, make possible, or accelerate? What is obsolesced by the artifact? What older, previously obsolesced ground is brought back and inheres in the new form? And finally, what will the new form reverse into when pushed to its extreme?

13 Ibid., 3.

14 D. Harlan Wilson, *Technologized Desire: Selfhood and the Body in Postcapitalist Science Fiction* (Hyattsville, Md.: Guide Dog Books, 2009).

15 Karen Barad, *Meeting the Universe Halfway: Quantum Physics and the Entanglement of Matter and Meaning* (Durham, N.C.: Duke University Press, 2007), 136.

16 Mark Poster, "The Information Empire," *Comparative Literature Studies* 41, no. 3 (2004): 318.

17 S. Herbrechter and I. Callus, "What's Wrong with Posthumanism?" *Rhizomes* 7 (2003), http://www.rhizomes.net/issue7/callus.htm.

18 Neil Badmington, "Introduction: Approaching Posthumanism," in *Posthumanism,* ed. Neil Badmington (New York: Palgrave, 2000), 9.

19 N. Katherine Hayles, *Writing Machines* (Cambridge, Mass.: MIT Press, 2002), 303.

20 McLuhan is again notable in this respect.

21 Johanna Drucker and Bethany Nowviskie, "Speculative Computing: Aesthetic Provocations in Humanities Computing," in *A Companion to Digital Humanities,* ed. S. Schreibman, R. Siemens, and J. Unsworth (Oxford: Blackwell, 2004), http://www.digitalhumanities.org/companion/; emphasis added.

22 Following Hansen's definition in *Bodies in Code,* 22, the adjective *medial* here marks "the specificity of analyses concerned with the materiality of the medium and of media generally."

23 Charles Mudede, "The Turntable," *CTheory* (2003), http://www.ctheory.net/articles/aspx?id=382.

24 To be clear, I mention this point to highlight a key difference between Mudede's project and my own rather than to mount a critique of his argumentation.

25 Antoine Hennion, "Music and Mediation: Toward a New Sociology of Music," in *The Cultural Study of Music: A Critical Introduction,* ed. M. Clayton, T. Herbert, and R. Middleton (London: Routledge, 2003), 84.

26 I use the term *praxis* in the sense that Agamben does in *What Is an Apparatus?,* 9—to indicate "a practical activity that must face a problem and a particular situation each and every time." Here the term indicates that

though posthumanism serves as a dominant critical lens through which I critique the artistic practices in question, the reverse is also the case.

27 Niklas Luhmann, as cited in Wolfe, *What Is Posthumanism?*, 231.

28 Bruno Latour, "Morality and Technology: The End of the Means," *Theory, Culture, and Society* 19, nos. 5–6 (2002): 248.

29 Dyens's theoretical work exists in the context of his notable efforts to compile (and creatively present) existing theories of posthumanism (often under other names, including "inhumanism," "transhumanism," and "humachinism"). These efforts doubly nominate him as a key figure of posthumanism.

30 Ollivier Dyens, *Metal and Flesh: The Evolution of Man: Technology Takes Over* (Cambridge, Mass.: MIT Press, 2001), 82. Note that cyborgs also function in Dyens's thought as literal instantiations of the "cultural bodies" that we all today possess. The cyborg metaphor does not, for Dyens, employ future technological advances (i.e., "brain cameras" and supersensitive hearing) to show where humans are heading but rather employs now obsolesced technologies (i.e., human flesh) to elucidate what we no longer are. Dyens's cyborg metaphor is, strictly, spoken from the perspective of a postcyborg.

31 For a distilled version of the type of advocacy for technological extensions of the human that characterizes transhumanism, see the World Transhumanist Association's "Transhumanist Declaration," available from http://transhumanism.org/index.php/WTA/declaration/.

32 Dyens, *Metal and Flesh*, 20.

33 As one example, Dyens argues in *Metal and Flesh*, 19–21, that standards of female beauty have remained fixed in proportion throughout our species's history and that this points to certain primal biological desires. What makes this assertion specifically scientific is not its subject (beauty) but the presumption that beauty constitutes a unified, testable, and falsifiable object of study, thus satisfying the criteria for scientific laws that McLuhan explains in *Laws of Media*, 3.

34 Judith Butler, "Competing Universalities," in *Contingency, Hegemony, Universality: Contemporary Dialogues on the Left*, ed. J. Butler, E. Laclau, and S. Žižek (New York: Verso, 2000), 157.

35 Hayles, *My Mother Was a Computer*, 3.

36 Cited in Stephen Ross, introduction to *Modernism and Theory* (New York: Routledge, 2009), 3.

37 Judith Butler, *Antigone's Claim: Kinship between Life and Death* (New York: Columbia University Press, 2000).

38 Hansen, *Bodies in Code*, 9.

39 The quality of "fieldness" is frequently found in McLuhan's accounts of acoustic space, where it refers to a space created by a set of relations (rather than being a physical container of relations). Richard Cavell cites McCaffery in *McLuhan in Space: A Cultural Geography* (Toronto, Ont.: University of Toronto Press, 2003), who notes that the dominant logic in such spaces is one of "event rather than . . . Euclidean [spatial geometry]" (157). This quality is often cited as a connection between acoustic space (as theorized by McLuhan) and contemporary theories of virtuality.

1. FROM GENES TO MEMES

1 In this chapter, the term *scientific* is used nominally. Emphatically, it is not intended to lumpenly reify all of science within the scope of this critique and should not be understood to refer to a universal "science." Instead, the term points back at the specific strains of scientific and evolutionary discourse that produce it in the sense that it is presented here.

2 Ollivier Dyens, *The Inhuman Continent: A Knowledge Interface* (2006), http://www.laconditioninhumaine.org/.

3 Ollivier Dyens, *Continent X* (1993), http://www.continentx.uqam.ca/.

4 John Doris, "Do You Know What You're Doing?" in *On the Human,* http://onthehuman.org/.

5 Ibid. Interestingly, Baudrillard seems to have anticipated these findings in *Paroxysm: Interviews with Phillippe Petit* (New York: Verso, 1998), 46, by arguing that "the will is always retrospective, [coming only] to sanction something that has already taken place. . . . You do something, and retrospectively, you conceive the plan."

6 Jacques Lacan, *The Seminar, Book XI, The Four Fundamental Concepts of Psychoanalysis,* trans. Alan Sheridan (1964; London: Hogarth Press and Institute of Psycho-Analysis, 1977), 232.

7 Richard Dawkins, *The Selfish Gene* (Oxford: Oxford University Press, 1976), 20.

8 While Dawkins's insistence on the relative importance—in Darwinian evolution—of genetic processes over those of individual species seems an obvious one, it is worth noting that it is a point that still today remains astonishingly unconsidered in popular North American culture, so completely ingrained is the notion of species mutation. Philosopher and aesthetician Denis Dutton, for example, is currently receiving significant acclaim for his book *The Art Instinct: Beauty, Pleasure, and Human Evolution* (New York: Bloomsbury Press, 2009), which includes a reading of Darwin that places sexual selection alongside (rather than subsidiary to) natural

selection, an argument that Dutton deems necessary to explain seemingly nonevolutionary decisions of humans. Although—as will become clear—I agree that the narrative of genetic evolution underdetermines certain excessive behaviors, the excessive character of these behaviors is precisely what precludes Dutton's teleological account. For a discussion of how this relates to artistic production, see the third chapter of Elizabeth Grosz's *Chaos, Territory, Art: Deleuze and the Framing of the Earth* (New York: Columbia University Press, 2008).

9 Dawkins, *Selfish Gene,* 190.

10 Ibid., 192.

11 Ibid., 193.

12 Ibid., 192. Ironically, the term *meme* has itself become a meme and today popularly signifies an idea or idiom that enters into common Internet usage. For example, idioms such as "ttyl" (meaning "talk to you later") are memes, as are popular stories, as are Internet actions such as "Rickrolling" someone (wherein a hyperlink is posted to an Internet message forum that appears to direct to a topical item but actually sends the user to a Rick Astley video on YouTube).

13 Ibid.

14 Joel Slayton, foreword to Dyens, *Metal and Flesh.*

15 Dyens is not, of course, alone in this observation. As Mark Hansen notes in "Media Theory," *Theory, Culture, and Society* 23, nos. 2–3: 299, for example, Bernard Stiegler follows paleontologist André Leroi-Gourhan in contending that human beings "evolve by passing on their knowledge through culture [such that they] are 'essentially' technical and have been so from their very 'origin.'" In order to differentiate it from strictly zoological evolution, Stiegler thus defines human evolution as irreducibly both biological and cultural; it occurs as a process that he dubs 'epiphylogenesis,' *evolution through means other than life,* [such that] the logic of the living [is contaminated] with the distinct and always concrete operation of technics."

In a different vein, Donna Haraway has also consistently and notably advocated for understanding semiotics and materiality as two aspects of the same thing, insisting in "Morphing in the Order: Flexible Strategies, Feminist Science Studies, and Primate Revisions," in *Primate Encounters,* ed. Shirley Strum and Linda Fedigan (Chicago: University of Chicago Press, 2000), 400, that "material-semiotic is one word for [her]." Finally, the notion of artificial life, a term coined by Christopher Langton, is predicated on this understanding, with M. Mitchell Waldrop quoting Langton, in *Complexity: The Emerging Science at the Edge of Order and Chaos* (New York: Simon and Schuster, 1992), 321, as noting that "evolution hasn't stopped.

It's still going on, exhibiting many of the same phenomena it did in biological history—except that now it's taking place on the social-cultural plane."

16 Dyens, *Metal and Flesh*, 15.

17 Ibid.

18 In *How We Became Posthuman: Virtual Bodies in Cybernetics, Literature, and Informatics* (Chicago: University of Chicago Press, 1999), 6, N. Katherine Hayles half-jokingly describes her "sleep agent" as competing with her "food agent" to dictate her next activity.

19 Dyens, *Metal and Flesh*, 10.

20 Arthur Kroker, *The Possessed Individual: Technology and the French Postmodern*, CultureTexts (Montreal, Quebec: New World Perspectives, 1992). Reversing Kroker's metaphor but amounting to a similar thing, Baudrillard argued in *Paroxysm*, 19, that "today, whether it be groups, nations or individuals, people are no longer fighting alienation but a kind of total dispossession."

21 Langdon Winner, *The Whale and the Reactor* (Chicago: University of Chicago Press, 1988).

22 Dyens, *Metal and Flesh*, 18. In the glossary of *Metal and Flesh*, 110, Dyens defines technological society as follows: "The era in which we currently live and within whose framework we conceive of technology as the ultimate idea, or the norm by which everything is defined, judged, and evaluated." He notes further that this connects to Neil Postman's definition of *technopoly* as a state of culture (and also a state of mind) that "consists in the deification of technology, which means that the culture seeks its authorization in technology, finds its satisfaction in technology, and takes its orders from technology" (110).

23 Ibid., 18.

24 Ibid., 10.

25 Ibid., 88–89.

26 Dyens thinks this question with greater nuance than this simplistic rendering suggests, specifically in *Metal and Flesh*, 45–46, when he considers Doyne Farmer's criteria for life. However, the specter of tautology remains active in his assertion that "living beings must be able to manipulate representations, for this is how they protect their biological integrity" (46). From this claim, for example, Dyens argues that viruses are included on the continuum of life, which contravenes the more conventional view—as Stephen Luper relates in *The Philosophy of Death* (Cambridge: Cambridge University Press, 2009)—that viruses are not "living things since they are neither organisms nor components of organisms" (14).

27 Stevan Harnad, "On Fodor on Darwin on Evolution," draft published

under Creative Commons Attribution No Derivatives (2009), 2. Fodor's lecture was given as part of the Hugh LeBlanc Lecture Series at the Université du Québec à Montréal, where Harnad serves as Canada Research Chair in cognitive sciences.

28 Ibid., 3.

29 Ibid., 2.

30 Dawkins, *Selfish Gene,* 4; emphasis mine.

31 This characterization draws on Baudrillard's description of the orders of simulation in *Simulations* (Los Angeles, Calif.: Semiotext(e), 1983), 83–92.

32 Arjun Appadurai, "Disjuncture and Difference in the Global Cultural Economy," *Theory, Culture, and Society* 7, nos. 2–3 (1990): 216–54.

33 Mark Poster, *What's the Matter with the Internet?* (Minneapolis: University of Minnesota Press, 2001). Poster charts the coevolution of certain cultural norms and print technologies. McLuhan also discussed this issue extensively throughout his oeuvre; see also "Flickering Connectivities in Shelley Jackson's *Patchwork Girl,*" in Hayles, *My Mother Was a Computer,* 143–67.

34 E.g., governments do not prohibit telepathic voyeurism because the laws of physics presumably preclude the need for them. Here again, McLuhan showed remarkable foresight in his prediction that electric communications would lead to a "discarnate man" that is constitutively immoral because he has no "natural law," in contrast to, for example, the "natural" laws of embodiment.

35 Especially in technologies of vision, which have dominated the medical and biotech industries (i.e., detailed medical exams, DNA analysis, synthetic biology).

36 Dyens, *Metal and Flesh,* 33.

37 Ibid.

38 Ibid., 19.

39 This problem can also be stated as follows: memes represent a detachment of processes of reproduction from biology, but biology is constitutive of these processes in the sense that it provides the lens through which a link between reproduction and survival is evinced. Thus memes require the very notion of genetic biology that they obsolesce.

40 To be clear, just because a technology is obsolesced does not mean that it ceases to be active. In fact, most cases of obsolescence simultaneously enhance an aspect of the technology that was not previously figured. Thus, for example, e-mail might be said to obsolesce letter writing as a form of communication, but this obsolescence endows letter writing with an increased quality of intimacy (and, often, formality) that it did not

previously possess. Thus, whereas certain technologies of letter writing are obsolesced by e-mail, others are (necessarily) enhanced; letter writing itself does not exist—technologically—except as a means of designating a field of relations that these other technologies enact.

41 Kroker, *Possessed Individual,* 13.

42 David Cecchetto and Émile Fromet de Rosnay, *Video Interview with Ollivier Dyens* (2009), http://davidcecchetto.net/research.html.

43 Transhumanists advocate for the technological extension of human capabilities. As such, the movement's symbol is H+, standing for "human enhancement," and might thereby be thought—as Cary Wolfe has suggested in *What Is Posthumanism?,* xv—as an intensification of humanism's predicates. Indeed, Nick Bostrom (a recognized leader of the movement) explicitly states that transhumanism "has its roots in secular humanism" in the official declaration of transhumanist values. See http://humanityplus.org/.

44 Dyens, *Metal and Flesh,* 19.

45 Ibid., 21.

46 Dyens cites Helena Cronin, a noted Darwinian philosopher and rationalist, who argued that standards of beauty are controlled by organic needs, noting that a waist that is 70 percent of the size of the hips is (1) a historically consistent standard of beauty and (2) a sign of optimal reproductive health (with respect to the subject's immune system and estrogen levels).

47 An analogous situation continues to play itself out in the field of cognitive psychology, where the discipline has frequently responded to P. J. Rushton's argument in "Race Differences in Behavior: A Review and Evolutionary Analysis," *Personality and Individual Differences* 9, no. 6 (1988): 1009–24, that race is a genetic determinant of IQ by critiquing his statistical data. See also Richard Nisbitt, *Intelligence and How to Get It: Why Schools and Cultures Count* (New York: W. W. Norton, 2009). Framed in this way, though, the debate neglects the possibility that IQ may actually measure a cultural construction of "whiteness" (for example) rather than (or in addition to) an "objective" value of intelligence (the quotation marks indicating, of course, that this objectivity has its own constitutive biases). If this were the case, then IQ tests would not yield measurements of an objective feature (such as intelligence) but would instead continually remake intelligence in the image of (culturally constructed and constantly shifting) racial boundaries.

48 This is not peculiar to Dawkins. As Simone de Beauvoir notes in *The Second Sex,* trans. H. M. Parshley (New York: Alfred A. Knopf, 1952), 10, "all physiologists and biologists use more or less finalistic language, if only because they ascribe meaning to vital phenomena."

49 Judith Butler, *Bodies That Matter: On the Discursive Limits of "Sex"* (New York: Routledge, 1993), 10.

50 Numerous spiritual perspectives disagree with this assumption: for example, a belief in reincarnation would at least reorient the discussion, as would many conceptions of an afterlife. In an entirely different register, the Derridean "trace" also poses significant challenges to the direct equation of survival and reproduction, as does Hayles's emphasis on pattern–randomness as supplemental to presence–absence (see chapter 3).

51 N. Katherine Hayles, "Unfinished Work: From Cyborg to Cognisphere," *Theory, Culture, and Society* 23, nos. 7–8 (2006): 163.

52 Dyens, *Metal and Flesh*, 52.

53 Slavoj Žižek, *The Sublime Object of Ideology* (New York: Verso, 1989), 32–33.

54 For an extended discussion of this understanding of ideological fantasy, see Slavoj Žižek's "How Did Marx Invent the Symptom," in *Sublime Object of Ideology*, 11–54.

55 Additional evidence that there is something at work in evolutionary theory that has not been accounted for can be found in the political history of the theory itself: what nominated evolution as the challenge to Christian values that it has today become (especially in the United States)? Geologists, after all, had already concluded a hundred years before *The Origin of Species* that the time scale of Genesis was wrong. This was the case not only in terms of when creation is alleged to have occurred but also with respect to the order and speed at which it occurred. And yet geology is hardly considered a controversial subject in the way that evolution is, suggesting an ideological—rather than a simply factual—difference between heritable genetic traits and a valued notion of evolution.

56 Isabelle Stengers gestures toward this point, arguing in *The Invention of Modern Science* (Minneapolis: University of Minnesota Press, 2000), 40, that "the actors in the history of the sciences are not humans 'in the service of truth,' if this truth must be defined by criteria that escape history, but humans 'in the service of history,' whose problem is to transform history, and to transform it *in such a way that their colleagues, but also those who, after them, will write history, are constrained to speak of their invention as a 'discovery' that others could have made.* The truth, then, is what succeeds in making history in accordance with this constraint."

57 "Badly analyzed composite" is a Bergsonian term that Gilles Deleuze takes up in *Bergsonism* (New York: Zone Books, 1988), 86, to claim that science is incapable of understanding the difference between multiplicity and the multiple and thus constitutively spatializes time. In this case, the claim is that scientific analysis collects qualitatively different concepts under a

single (quantitative) lens; specific to evolution, Deleuze cites Bergson to claim that its mistake "is to conceive of vital variations as so many actual determinations that should then combine on a single line" (99), in this case, survival.

58 Bruno Latour has convincingly made this point in numerous contexts. For example, he recently noted in "'It's Development Stupid!' Or: How to Modernize Modernization," *RETHINK: Contemporary Art and Climate Change* (2009), http://www.rethinkclimate.org/, that, in the traditional narrative of "Progress," "Science (capital S) is the shibboleth that defines the right direction of the arrow of time because it, and it only, is able to cut into two well separated parts what had remained in the past hopelessly confused: a morass of ideology, emotions and values on the one hand, and, on the other, stark and naked matters of fact." The point, for Latour, is that science has thus tended to detach the entangled entities of this imbroglio, thereby producing its narrative of emancipation; this point is also well made in Bruno Latour, *Pandora's Hope: Essays on the Reality of Science Studies* (Cambridge, Mass.: Harvard University Press, 1999).

59 Dyens, *Metal and Flesh*, 62.

60 Ibid., 88.

61 As Hayles notes in *My Mother Was a Computer*, 26–27, the arrow of emergence in complex systems tends to move in a single direction.

62 In this light, it is notable that complexity only really started to bear fruit for scientists once it was measurable via a "power law distribution," which indicates that the likelihood of a given event is inversely proportional to the size of its disruption to the system. Prior to this, it was not clear how one could tell, precisely, what was simple and what was complex. As Waldrop, *Complexity*, 308, notes, though, with power law distributions, "you can tell that a system is at the critical state and/or the edge of chaos [i.e., where emergence takes place] if it shows waves of change and upheaval on all scales *and if the size of the changes follows a power law*" (emphasis mine). Notably, these power laws were themselves drawn from existing notions of organization and disruption—initially, earthquake fault lines—where the distinction between organization and chaos is more clearly related to empirical evidence than it is in Dyens's application.

63 Dyens, *Metal and Flesh*, 27.

64 Kroker, *Possessed Individual*, 120.

65 To be clear, *conflation* here does not mean "misrepresentation" or "misattribution."

66 Philip Mirowski, "Book Review: 'Metal and Flesh,' by Ollivier Dyens," *The Information Society* 20, no. 65 (2004): 65.

67 Mirowski, *ISIS* 91, no. 3 (2000): 639–40, offers a more qualified, but no less scathing, review of Hayles's *How We Became Posthuman*, similarly citing a lack of substantial scientific knowledge. In Hayles's case, her significant (but relatively understated) scientific training further suggests that critiques that are grounded in claims to disciplinary privilege (such as Mirowski's) may belie other unspoken agendas and presumptions.

68 Wolfe, *What Is Posthumanism?*, 43.

69 Dawkins, *Selfish Gene*, ix.

70 N. Katherine Hayles, "Desiring Agency: Limiting Metaphors and Enabling Constraints in Dawkins and Deleuze / Guattari," in *SubStance* 30, nos. 1–2 (2001): 144.

71 Nicholas Maxwell asks this question (under the alibi of a relatively conservative call for "wisdom" in science) in *Is Science Neurotic?* (London: Imperial College Press, 2004), arguing that empiricism acts as a metaphysical assumption of scientific inquiry that ultimately leads to irrationality.

72 In this context, it is a useful corrective to note—as Stephen J. Gould does in *Rocks of Ages: Science and Religion in the Fullness of Life* (New York: Ballantine Books, 1999)—that no fewer than three popes have acknowledged evolution, each insisting that it does not pose a theological problem for Roman Catholicism.

73 In chapter 5, I argue that this remains the case even in Hansen's reading of autopoietic theory, through which he attempts to structure the viewer's contingency into his claims.

74 Luhmann, as cited in Wolfe, *What Is Posthumanism?*, 122. All this points to a more general question with respect to Dyens's thought: is it that acquisition of knowledge—in this case scientific knowledge—makes visible a broader portion of a spectrum that exists independent of its perception, or is it instead that the perspective from which knowledge appears constitutes not only the knowledge but also the particular spectrum that that knowledge exists within? It is telling that Dyens seems to occupy both answers: on one hand, his insistence on multiply constituted realities bearing no relation to an extradiscursive Real claims the latter; and yet, when asked this very question in an interview, Dyens identifies himself as a "profoundly materialist" person who, for practical purposes, believes the former. See Cecchetto and Fromet de Rosnay, *Video Interview with Ollivier Dyens*.

75 Similarly, Dyens's most recent book, *La Condition inhumaine, essai sur l'effoi technologique* (Paris: Flammarion, 2008), not yet translated, is published by perhaps the leading French intellectual press.

76 Ollivier Dyens, *Living in the Inhuman Condition: The World Is Technology*, Lansdowne Lecture Series (Victoria, B.C.: University of Victoria, 2009).

77 Dyens offered statistical evidence of the remarkable effectiveness of join-ing behavioral predictors with marketing techniques in ibid.

78 Marshall McLuhan, "Myth and Mass Media," *Daedalus* 88, no. 2 (1959): 340.

79 Ibid., 346.

80 Negative replicators follow from the positive definition of life that I have been unpacking. In essence, Dawkins, *Selfish Gene,* notes that any gene that does not express a particular characteristic can be described as a gene that expresses not having that characteristic. Thus, for example, most cats share the gene for not being polydactyl (90).

81 Baudrillard, *Paroxysm,* 25–38.

82 Ibid., 65.

83 Ray Kurzweil predicts in *The Singularity Is Near: When Humans Transcend Biology* (New York: Penguin, 2006) that, by 2040, the rate of medical tech-nologies' development will exceed the rate of human deterioration due to aging and disease, which is to say that life expectancy will increase beyond a 1:1 ratio with time. As a result, although individual humans will not be immortal, they will be able to expect to live forever.

84 Indeed, this is not only the case genetically but also with respect to claims to reasoning. As Sarah Blaffer Hrdy has recently argued in a post titled "How Humans Became Such Other-Regarding Apes" (2009) in the "On the Human" forum, "our closest relatives among the other apes, chimpan-zees and bonobos, with whom we last shared common ancestors some seven million years ago, and still share nearly 99% of DNA sequences, also descend from highly social, manipulative ancestors and possess simi-lar cognitive capacities, yet they are far more single-mindedly self-serv-ing. In this respect, other apes are far more nearly 'rational actors' than humans are."

2. DARK MATTERS

1 This is a theme that is raised throughout McLuhan's writing. Visual space is characterized by such features as rationalism, specialism, objectivity, and the detachment of figures from their ground. See *Laws of Media,* 204–5, for two of McLuhan's "visual space" tetrads.

2 In *The Parasite* (Minneapolis: University of Minnesota Press, 2007), 38, Serres notes that to play the position is to have a relation "with the relation itself" rather than to contents as "sources of relations." This chapter of the book seeks in part to unpack this assertion as it relates to sound; see also chapter 6.

3   Limited documentation—including images, an audio sample, links to artists' and curators' talks, and the exhibition monograph—is available via http://www.davidcecchetto.net/artistic.html.

4   Cary Wolfe, "'Bring the Noise': The Parasite and the Multiple Genealogies of Posthumanism," in Serres, *Parasite*, xxiii.

5   Serres, *Parasite*, 79. Wolfe notes this passage, and provides a larger context for it, in his introduction to the new edition of *Parasite*, xii–xiv.

6   A number of children attended the exhibit, many of whom were shorter than the height of the suspended panels, adding a slightly grotesque perspective to the reading I am offering, namely, children trapped in a basement that unilaterally frames their experience. In this, conventional "childish" fears of ghosts manifest precisely as a foreclosure of haunting, the ultimate horror of inevitability.

7   Mark B. N. Hansen, *New Philosophy for New Media* (Cambridge, Mass.: MIT Press, 2004), 299.

8   Hayles, "Desiring Agency," 148.

9   Ibid., 150.

10 Ibid., 151.

11 See chapter 1 for a more extended discussion of Dawkins's selfish gene theory.

12 Wolfe, *What Is Posthumanism?*, 133.

13 The syntax of this paragraph, though not its content, is borrowed from the introduction to Butler's *Bodies That Matter*.

14 For a discussion of autonomic consciousness, see Ted Hiebert, *In Praise of Nonsense: Aesthetics, Uncertainty, and Postmodern Identity* (Montreal: McGill Queens University Press, 2012), 156–60.

3. N. KATHERINE HAYLES AND HUMANIST TECHNOLOGICAL POSTHUMANISM

1   Though I arrived at the term independently, my reading aligns with the notion of "humanist posthumanism" developed by Wolfe in *What Is Posthumanism?*, indicating an "*internal* disciplinarity [that remains] humanist through and through," despite a theorist "taking seriously the existence of nonhuman subjects" in her conception of a "discipline's *external* relations to its larger environment" (123–24).

2   Hayles, *How We Became Posthuman*, 2.

3   Ibid.

4   In *Parables for the Virtual: Movement, Affect, Sensation* (Durham, N.C.: Duke University Press, 2002), 10, Massumi speaks of retroduction as "a production, by feedback, of new movements [such that] a dynamic unity has been

retrospectively captured and qualitatively converted. [For example] space itself is a retroduction, by means of the standardization of measurement."

5   Importantly, Hayles differentiates the specificity of embodiment from the normatively operative notion of "the body." See, for example, her discussion of Foucault in chapter 8 of *How We Became Posthuman*. The importance of this differentiation is unpacked more fully in chapter 5 of this book.

6   Hayles, *How We Became Posthuman*, 54.

7   As we will see later, the presence–absence of this being exists symbiotically with the pattern–randomness of information.

8   Hayles, *How We Became Posthuman*, 4–5.

9   Ibid., 22.

10 Ibid., 7.

11 Ibid., 66.

12 "Autopoietic reflexive systems" are systems that self-generate independent of any conscious will.

13 Hayles, *How We Became Posthuman*, 136.

14 Ibid.

15 Humberto Maturana and Francisco Varela, *Autopoiesis and Cognition: The Realization of the Living* (Dordecht, Netherlands: D. Reidel, 1980), 80.

16 Hayles, *How We Became Posthuman*, 155.

17 Ibid., 158.

18 Though Hayles and Varela are both proponents of "embodied philosophy," their projects—that is, what they mobilize their perspectives in service of—remain quite different. Because the critique that I offer of Hayles in this chapter focuses primarily on her philosophical project, it is not intended to extend to a critique of Varela.

19 N. Katherine Hayles, "Unfinished Work: From Cyborg to Cognisphere," *Theory, Culture and Society* 23, nos. 7–8 (2006): 165.

20 Hayles, *How We Became Posthuman*, 13–14.

21 Hayles, ibid., 17, defines a skeuomorph as "a design feature that is no longer functional in itself but that refers back to a feature that was functional at an earlier time."

22 Hayles, *My Mother Was a Computer*, 18.

23 Ibid., 4.

24 Cited in ibid., 19.

25 Ibid.

26 Ibid., 20.

27 Ibid., 3.

28 Hayles, "Unfinished Work," 163. Interestingly, many of the predicates of

the RoC can be found (without the significant cultural support of ubiquitous computers) much earlier. As Hannah Arendt, *The Life of the Mind* (New York: Harcourt, 1971), 59, notes, for example, the "conviction that mathematical reasoning should serve as a paradigm for all thought is probably as old as Pythagoras; at any rate we find it in Plato's refusal to admit anyone to philosophy who has not been trained in mathematics." For Arendt, this conviction (combined with premodern events such as the reformation, the discovery of the "New World," and the development of the telescope) is part and parcel of "world alienation" (i.e., where the world is viewed from a transcendent—specifically scientistic or mathematically based—perspective), an argument that in many ways prefigures Hayles's perspective.

29 Hayles, *My Mother Was a Computer*, 219.

30 Ibid.

31 Žižek, *Sublime Object*, 6.

32 Slavoj Žižek, *The Parallax View* (Cambridge, Mass.: MIT Press), 17.

33 Žižek, *Sublime Object*, 158. The mark of this nonsignifying reference is the *objet petit à*, "the original lost object which in a way coincides with its own loss." Ibid.

34 Hayles positions this in contrast to Žižek, but it is arguable whether she actually attains the distance from him that she claims. For example, her analysis, in *My Mother Was a Computer*, of the novel *Permutation City* concludes that the narrative "implies that computation has been transformed from a metaphor into the means by which reality is generated, a means that includes the illusions of those," such as Žižek, in Hayles's reading, who think the UC is (only) a metaphor (224). This may not be an accurate reading of Žižek, though, as he frequently emphasizes that ideology is precisely not metaphoric but rather operates as the material instantiation of the deep-seated trauma that *is* the Real. This disputation bears noting because it exemplifies Hayles's broader tendency—especially in her earlier work—to gain political traction for a perspective by working it against a questionable reading of an established theorist. This tendency is discussed further at the close of this chapter.

35 Hayles, *My Mother Was a Computer*, 23.

36 Vis-à-vis McLuhan, computation might be said to resemble nonphonetic language in this respect.

37 In *My Mother Was a Computer*, 47, Hayles ventures to guess that, in contemporary critical theory, *signifier* is used thousands of times more than *signified*. Moreover, despite Derrida's insistence on their co-implication, numerous critics have followed Bruno Latour's insistence in *We Have Never*

Been Modern (Cambridge, Mass.: Harvard University Press, 1993), 6, that deconstruction grasps the specificities of practice "as badly as possible."

38 Again, this is key to deconstruction in general. Derrida's project was not only to dismantle metaphysics by showing its absent center but also to begin to write the ghostly presences—the hauntologies—of metaphysics that are beyond the pale of positivistic language.

39 Hayles, *My Mother Was a Computer*, 7.

40 The contrast between emergence and intermediation is a finer one than is suggested here, notably because the former *does* include both initial and emergent agents in subsequent orders of emergence. The distinction obtains, though, because complexity continues to function as a metric in emergence so that emergent behaviors are always more complex than their provenance would suggest. Intermediation, by contrast, is not concerned with a metric of complexity but rather with the different qualities of the relations that emerge.

41 Hayles, *My Mother Was a Computer*, 39.

42 Ibid., 8.

43 Ibid., 40.

44 Ibid., 41.

45 John Johnston makes a related argument in "Machinic Vision," *Critical Inquiry* 26 (Autumn, 2009): 27, where he positions machinic vision as presupposing "not only an environment of interacting machines and human-machine systems but a field of decoded perceptions that, whether or not produced by or issuing from these machines, assume their full intelligibility only in relation to them."

46 Hayles, *My Mother Was a Computer*, 50.

47 Ibid., 44.

48 Ibid., 48.

49 Ibid.

50 Ibid., 249.

51 Ibid., 66.

52 Ibid., 43.

53 Hayles spends a significant portion of *How We Became Posthuman* making a similar point, showing that pattern is always infused with a degree of randomness. As with the resulting semiotic square in that account, the significance of the term noise here is that it moves from a paradoxical duality (i.e., presence–absence) to an active set of relations.

54 Hayles, *My Mother Was a Computer*, 68.

55 Ibid.

56 Ibid.

57 Ibid., 50.

58 This is a fitting site of analysis, considering that intermediation itself serves a narrative function for Hayles (i.e., in her writing).

59 Hereinafter, all references to "database" in this chapter refer specifically to relational databases. The novelty of relational (as opposed to hierarchical) databases is, in short, that data in a relational database can be stored, added to, or manipulated without impacting other elements in the database. Similarly, relational databases can be queried without specific prior knowledge of the databases' content.

60 N. Katherine Hayles, "Narrative and Database: Natural Symbionts," *PMLA* 122, no. 5 (2007): 1605.

61 In this respect, Dawkins's selfish gene argument exemplifies database thinking in a biological register.

62 Hayles, "Narrative and Database," 1603.

63 Hayles, *My Mother Was a Computer,* 31.

64 Lev Manovich, *The Language of New Media* (Cambridge, Mass.: MIT Press, 2001), 227.

65 Ibid., 231. Manovich contrasts this to language (particularly to Saussure via Barthes), where "elements in the syntagmatic dimension are related *in praesentia,* while elements in the paradigmatic dimension are related *in absentia*" (230).

66 Ibid., 227.

67 Hayles, "Narrative and Database," 1605.

68 Ibid., 1604–5.

69 Ibid., 1603.

70 Hayles, "Unfinished Work," 160.

71 Hayles, *How We Became Posthuman,* 3.

72 In this context, "economy of scarcity" refers to a unified conception of the self that can be possessed, exchanged, or dispossessed, but only through binary operations. For example, if a "self" is totally possessed by genetic forces (as in Dawkins), it cannot claim to also possess itself agentially.

73 Hayles, *My Mother Was a Computer,* 63.

74 Hayles, *How We Became Posthuman,* 3.

75 Ibid., 3–4.

76 Similarly, Hayles maintains the fundamental belief that if our bodies were differently constructed, our sense of logic would also be altered. From this, she observes that our bodies *are* different from one another (and from our ancestors' bodies), from which she develops—in *How We Became Posthuman,* 192–221—a dialectic between the (normative) body and embodiment.

77 Bill Viola, "Will There Be Condominiums in Data Space?," in *Reasons for Knocking on an Empty House*, ed. R. Violette (Cambridge, Mass.: MIT Press, 1995), 123.

78 Hayles convincingly illustrates this point in a discussion of Radio Frequency Identification (RFID) technologies in "RFID: Human Agency and Meaning in Information-Intensive Environments," *Theory, Culture, and Society* 26, nos. 2–3: 55, where she notes that "RFID participates in a paradigm shift in which the focus shifts from present and past actions to the anticipation of future actions," a change that limns the different temporality at play in an RFID ecology.

79 Hayles, *My Mother Was a Computer*, 200.

80 Richard Wray, "French Anti-filesharing Law Overturned," *The Guardian*, June 10, 2009.

81 Hansen, *Bodies in Code*, 39; emphasis added.

82 Hayles, *My Mother Was a Computer*, 239.

83 Galloway, as cited in Hayles, "RFID," 53.

84 Hayles, *My Mother Was a Computer*, 239.

85 In fact, it is unlikely that Žižek would disagree with this line of critique as he regularly argues that historicization has—in much of contemporary theory—gone too far in neglecting the basic ahistorical conditions of subjectivity. That is, Žižek follows Lacan in the belief that the unconscious is structured like a language and the human subject is born into systems of meaning; as a result, while the systems themselves are not fixed, Žižek asserts that being born into a system (or systems) is an ahistorical fact.

86 Guattari, as cited in Hayles, *My Mother Was a Computer*, 176.

87 Žižek, *Sublime Object*, 165.

88 Žižek, *Parallax View*, 7.

89 Hayles, *My Mother Was a Computer*, 175.

90 Ibid., 211.

91 I.e., Hayles (along with a host of thinkers ranging from technologists to theorists to policy makers) notes that privilege manifests as possession (ranging from literal ownership of goods to self-possession) in the economy of scarcity that is constitutive of analog relations (i.e., if person A owns a particular table, person B does not). By contrast, since there is no "original" in a digital economy, privilege becomes a matter of gaining access (i.e., if person A has unfettered access to the hard drive of person B, he effectively has access to the privilege afforded by that information). Clearly these worldviews are not distinct (as Hayles notes), and Benjamin's well-known reading of film—in its predigital manifestation—in "The Work of Art in the Age of Mechanical Reproduction," in *Selected Writings*

of Walter Benjamin, vol. 4, *1938–1940*, ed. H. Eiland and M.W. Jennings, 251–83 (Cambridge, Mass.: Belknap Press of Harvard University Press, 2003), is also regularly cited in this respect.

92 Hayles, *My Mother Was a Computer*, 175.

93 Ibid., 172.

94 Hayles, "Desiring Agency," 147.

95 Citing Dene Grigar, Hayles proposes in *My Mother Was a Computer*, 89, that "the adage that something is gained as well as lost in translation applies with special force to print documents that are imported [i.e., translated] to the Web."

96 Hayles, *How We Became Posthuman*, 4.

97 Hayles, *My Mother Was a Computer*, 192.

98 Ibid., 177. Nicholas Gane notes in "Radical Post-humanism: Friedrich Kittler and the Primacy of Technology," *Theory, Culture, and Society* 22, no. 3 (2005): 25–41, the influence of Lacan on Kittler, in particular, the former's 1954–55 seminars on "The Ego in Freud's Theory and in the Technique of Psychoanalysis." At stake is a restaging of the information theory of Shannon and Weaver: "Lacan's theory of the ego is formulated through direct engagement with early cybernetic theory, and makes reference to Shannon . . . and also Norbert Wiener. . . . Lacan's writings on this subject have titles such as 'Homeostasis and Insistence,' 'Freud, Hegel and the Machine,' 'The Circuit,' and 'Psychoanalysis and Cybernetics' (the title of his lecture to the Société Française de Psychanalyse delivered on 22 June 1955)" (32).

99 Ibid.

100 Ibid., 191.

101 Ibid., 173.

102 Hayles, *How We Became Posthuman*, 3–4.

103 Hayles, "Unfinished Work," 161.

104 In *The Five Senses: A Philosophy of Mingled Bodies* (New York: Continuum, 2008), 25, Serres evocatively states that "all dualism does is reveal a ghost facing a skeleton."

105 In *My Mother Was a Computer*, 60–61, Hayles cites Wendy Hui Krong Chung, who argues that "*software is ideology*, instancing Althusser's definition of ideology as 'the representation of the subject's imaginary relationship to his or her real conditions of existence.'"

106 Ibid., 146.

107 The relation between ethics and politics is, of course, a highly contested discourse. In Hayles's case, I would suggest that her emphasis on intermediation promotes (or perhaps assumes) a view wherein the two are

relatively continuous. The theoretical issues that stem from this are part and parcel of the critique that follows below.

108 Jack Reynolds, "Jacques Derrida (1930–2004)," in *Internet Encyclopedia of Philosophy: A Peer-Reviewed Academic Resource* (2010), http://www.iep.utm.edu/derrida/.

109 Hayles, *My Mother Was a Computer,* 93.

110 Wolfe, *What Is Posthumanism?,* 99.

111 Video documentation of this 2008 lecture, titled "Narrating Consciousness: Language, Media, and Embodiment" (including the question-and-answer period), is available from http://www.pactac.net/pactacweb/web-content/video66.html.

112 Ibid.

113 Ibid.

114 This is ironic, coming from Hayles, because it is a line of critique that she frequently levels at other theorists. For example, in "Desiring Agency," she argues that Deleuze and Guattari share with Dawkins "a certain effacement of the linguistic actors they rely on to perform what [their texts describe]" (156).

115 Hayles, *My Mother Was a Computer,* 3.

116 Ibid., 204. As is briefly discussed later, Hayles's use of "we" in such formulations also suggests a certain ultimate autonomy that is afforded to humans in her schema.

117 Ibid., 89.

118 Ibid., 3.

119 Baudrillard makes this point well in *Paroxysm,* 38, noting that "the problem is how, at the heart of [the definitive indeterminacy of modern science] laws can appear and the reality effect can emerge. This is where the problem turns around. It isn't the nothing—the other of the real, the other of rationality—which is a problem, but the real itself." Similarly (though from a completely different theoretical perspective), Theodor Adorno and Max Horkheimer's unfolding of the "irrationality of reason" in *Dialectic of Enlightenment,* trans. Edmund Jephcott (Stanford, Calif.: Stanford University Press, 2002), is apropos to this argument.

120 Hayles, *My Mother Was a Computer,* 56.

121 Ibid., 197.

122 Ibid., 208.

123 Hayles, "Narrating Consciousness."

124 Hayles, *My Mother Was a Computer,* 205.

125 Baudrillard, as cited in Steven Poole, "Meet the David Bowie of Philosophy," *The Guardian,* March 14, 2000.

126 In *My Mother Was a Computer*, 197, Hayles relates autism to the superintelligence of computers, noting "autistic people have no model in their minds for how others act." While this leads to their perceiving "most actions as inexplicable and frightening," it also allows them to organize data at a speed that eludes non-autistic people.

127 Ibid., 48–49.

128 Ibid., 242.

129 Tim Lenoir, "(Foreword) Haptic Vision: Computation, Media, and Embodiment in Mark Hansen's New Phenomenology," in *New Philosophy for New Media* (Cambridge, Mass.: MIT Press, 2004), xix.

130 Hayles, *How We Became Posthuman*, 81. Janet Freed was a conference assistant at the Macy Conferences who was mislabeled "Janet Freud" in a documentary photo of the events.

131 Hayles, *My Mother Was a Computer*, 190. It should be noted that this is not just a single occurrence. Much of *How We Became Posthuman* deals with the necessity of supplementing what Hayles identifies as the Lacanian dialectic of presence/absence (30) in light of the issues brought to bear by information technologies. In addition, Hayles's reading of texts as assemblages in *My Mother Was a Computer* is an explicit reference to Deleuze and Guattari, a fact made even more significant to my argument here in light of her claim that Guattari understates his own theoretical proximity to Lacan (176).

132 Ibid., 216.

133 Hayles, *How We Became Posthuman*, 22.

134 For an examination of the governmentality of a politics of inclusion as it operates through tolerance, see Wendy Brown's deconstruction of the discourse of tolerance in *Regulating Aversion: Tolerance in the Age of Identity and Empire* (Princeton, N.J.: Princeton University Press, 2006).

4. THE TRACE

1 This description paraphrases that offered by Lozano-Hemmer in the video documentation for *The Trace*; see http://www.lozano-hemmer.com/.

2 I use the term *locative* here, rather than *tactile* or *embodied*, to indicate that the work's specificity comes about through a projection of the participants' location rather than through an attempt to address their internal biological sensations (as in, e.g., media artist Char Davies's *Osmose* and *Éphémère*).

3 Jay Garmon notes in "Geek Trivia: First Shots Fired," *Tech Republic*, May 24, 2005, http://www.techrepublic.com/article/geek-trivia-first-shots-

fired/5710539?tag=content;siu-container, that the first-person shooter genre of video game, for example, rose to prominence in the early 1990s with games such as Wolfenstein 3D (1992) and Doom (1993). These games were not the earliest examples of this technology but were prominent in its dissemination. Nonetheless, it bears noting that Jeffrey Shaw's *The Legible City* (1989) combined the first-person shooter graphical model with a bicycle interface and existing architectural structures prior to these games' release.

4   Christiane Paul argues in *Digital Art* (New York: Thames and Hudson, 2003), 96, that "the creation of a believable world requires continuity: the environment needs to develop in a continuous way."

5   I.e., the visual graphics literally move through space, whereas sound gives the illusion of a sound source that moves through space. In actuality, the sound itself is distributed according to relative intensities (see the introduction to this book for a short discussion of the physics of sound in this respect).

6   E.g., if the local participant moves ten centimeters across and twelve centimeters vertically, this might be graphically rendered as a ten-centimeter horizontal movement and a 1 percent vertical movement (because the graphic's vertical space is depicted through simulated shadows). Sonically, though, the action would be rendered as a ten-centimeter horizontal and twelve-centimeter vertical movement (1 percent of the actual space rather than 1 percent of the simulated depth). In both cases, this difference is compounded by the local participant not knowing whether the remote space is precisely the same size as the local station so that he does not know to what extent the telepresent movements align with the remote participant's actual movements in space. As a result, the relation between actions and their representations differs between media (because they use different base metrics) and potentially between locations (because they may be different sizes). Moreover, the fact that the computer treats the two stations as though they are exactly the same size (despite this being impossible to achieve perfectly) actualizes this potential difference in a way that is different from how it might be imagined by the participants.

7   In *Bodies in Code*, 37, Hansen notes a similar feature in Mylon Kruegers's *Videoplace*, insisting that "by coupling the motile body with graphic elements that do not visually imitate or simulate it, *Videoplace* opens a disjunction between the body image and the body schema" that makes the latter *sensible*.

8   Hansen, *Bodies in Code*, 94.

9   This incompatibility points to the excess of code to language that was discussed, vis-à-vis Hayles, in chapter 3.

10 http://www.lozano-hemmer.com/.

11 Ibid.

12 Remembering the discussion of forced choice in chapter 3, it bears emphasis that this does not mean that the participants have full control of their actions but only that—in contrast to a liberal humanist perspective, but in total agreement with Lacan—they are born into a system of meaning that exceeds them.

13 This consciousness is "ethical" in that it pertains to the participants' codes of relating to one another.

14 This is altogether typical of social digital media such that Mary Bryson has argued in, "Can We Play 'Fun Gay'? Disjuncture and Difference in Millennial Queer Youth Narratives," in *Critical Digital Studies Workshop* (Victoria, B.C.: University of Victoria, 2009), http://www.pactac.net/pactacweb/web-content/video77.html, that "socially networked media mitigate against the very possibility of a robust relationality of anonymous strangers. In the networked economy . . . whatever, or whomever, cannot be [identified is relegated to the periphery and deemed suspect]."

15 Judith Butler, *The Psychic Life of Power* (Stanford, Calif.: Stanford University Press, 1997), 10.

16 Ibid.

17 As cited in chapter 3, Hayles argues in *How We Became Posthuman*, 4, that one of the shared features of most strains of posthumanism is the notion that there is "no *a priori* way to identify a self-will that can be clearly distinguished from an other will."

18 Judith Butler, *Precarious Life: The Powers of Mourning and Violence* (New York: Verso, 2006), offers the situations of Palestinian and homosexual lives in North America as an example.

19 Butler, *Antigone's Claim*, 70.

20 E.g., in ibid., Butler considers the status of the father for a child of a single mother: is the father still there as a spectral "position" or "place" that remains unfilled, or is there no such "place" or "position"? To wit, she asks, "Is the father absent, or does the child have no father, no position, and no inhabitant? Is this a loss, which assumes the unfulfilled norm, or is it another configuration of primary attachment whose primary loss is not to have a language in which to articulate its terms?" (69).

21 To be clear, this critique does not point to particular instances of sexuality, gender, and desire but rather to the categories—that is, the "forms"—themselves.

22 This is particularly true with respect to the way in which ambivalence is mobilized in connection with grief in Butler, *Precarious Life*. That is, Butler registers grief as a certain mode of being outside oneself (one is "beside oneself" with grief) so that it necessarily involves agreeing to undergo a transformation, "the full result of which one cannot know in advance" (21). To grieve, then, is literally to make oneself vulnerable to a certain sense of destiny, to something that is larger than "one's own knowing and choosing" (21); to grieve is to be taken hold of by something, a position that is (paradoxically) both chosen (in that we make ourselves open to grieving) and imposed (in that grieving involves the recognition that a part of oneself has already been lost in the loss being grieved).

23 Moreover, this primary vulnerability—which is foundational to the participants' actions—suggests that their subjectivity is retroactively produced in the same sense that analog subjectivity is.

24 I owe the formulation of this sentence to Wolfe's discussion of the character Selma in Lars von Trier's film *Dancer in the Dark* in *What Is Posthumanism?*, 148.

25 Gane, "Radical Post-humanism," 40.

26 In *Life of the Mind*, 57, Hannah Arendt differentiates cognition and reason, via Kant, as the difference between apprehension and comprehension: the former seeks to grasp what is "given to the senses," and the latter seeks to understand its meaning.

27 Butler, *Psychic Life of Power*, 211.

28 *Value-form* here indicates the form in which an entity takes on value, which is to say—following Butler—takes on meaning (broadly construed).

29 Butler notes in *Psychic Life of Power*, 17–18, that the subject exceeds the law of noncontradiction but remains bound to it as the condition of its intelligibility.

30 Paul Virilio, "Big Optics," in *On Justifying the Hypothetical Nature of Art and the Non-identicality within the Object World*, ed. P. Weibel (Cologne, Germany: Galerie Tanja Grunert, 1992), 93.

31 McLuhan, *Laws of Media*, 100.

5. FROM AFFECT TO AFFECTIVITY

1 Hansen's work has garnered significant and sustained attention since the publication of *New Philosophy for New Media*, particularly in the North American vein of media studies and in the emerging discourses of posthumanism. Compelling in its own right, Hansen's work is also a valuable site of inquiry because it combines frequently cited theories of affectivity and

systems theory, each of which has been highly influential in the developing discourse(s) of posthumanism (as well as in the parent disciplines on which posthumanism draws).

2 N. Katherine Hayles, "Clearing the Ground (Foreword)," in Hansen, *Embodying Technesis: Technology beyond Writing* (Ann Arbor: University of Michigan Press, 2000), 7.

3 Hansen, *Bodies in Code*, 9.

4 Hayles, "Clearing the Ground," v.

5 Hansen uses *translation* in the sense that Bruno Latour does in *Science in Action: How to Follow Scientists and Engineers through Society* (Cambridge, Mass.: Harvard University Press, 1997), 132–44.

6 Hansen, *Embodying Technesis*, 1–2.

7 Ibid., 2–3.

8 Ibid., 2.

9 Ibid., 3.

10 Ibid.

11 Ibid., 93.

12 Ibid., 4. In this chapter, I use the term *representation* in the broad sense indicated by Hansen, where he notes that by a "representationalist approach [he means] any approach that legitimizes representation as its frame of reference, whether for ends either affirmative or critical, positivist or deconstructive" (ibid.).

13 Karen Barad, "Posthumanist Performativity: Toward an Understanding of How Matter Comes to Matter," *Signs: Journal of Women in Culture and Society* 28, no. 3 (2003): 801.

14 Hansen, *Embodying Technesis*, 20.

15 Cited in ibid., 86–87.

16 Ibid., 80.

17 Ibid., 79. In particular, this criticism is directed toward the introduction of second-order cybernetics into systems theory, where the former insists that any observation that introduces a distinction is itself "unable to observe the distinction on which it bases its own observation." Schwanitz, as cited in ibid., 79. By thus shifting the focus of analysis from a "first-order observing of objects to a second-order observation *of observations*," Hansen insists that systems theory radically isolates system from environment, thereby "cutting description off from embodied reality" (79–80).

18 Ibid., 78.

19 Ibid., 14.

20 Ibid., 86.

21 Ibid., 84.

22 Ibid., 82.

23 Ibid., 83.

24 Ibid., 85.

25 Jacques Derrida, *Of Grammatology* (Baltimore: Johns Hopkins University Press, 1974), 158.

26 Ibid., 159.

27 Hansen, *Embodying Technesis,* 14.

28 To be clear, the Derridean claim is not necessarily that all technical appa-ratuses partake of language's instability and ambiguity but rather that our knowledge of them does (and that they come to be, for us, only through becoming objects of knowledge). As suggested earlier, though, the inverse is also true in that Derridean claims about language are predicated on language's *not* being understood as a system that is closed off from the operations of technics. Thus the issue is not whether biological processes can operate independently of language but rather that extrapolating these processes into an observational register (or a predictive or categorical one, for that matter) to delve into their meaning is not a neutral endeavor: understanding digestion as a digestive process, rather than as a random or even stochastic set of events, necessarily presumes a frame of reference (i.e., an individual or a part of an individual) that is privileged linguistically.

29 Derrida, as cited in Hansen, *Embodying Technesis,* 85.

30 Ibid.

31 Derrida, *Of Grammatology,* 61.

32 Gayatri Chakravorty Spivak, translator's preface to Derrida, *Of Gram-matology,* xvii.

33 Jacques Derrida, "Signature Event Context," in *Limited Inc.* (Evanston, Ill.: Northwestern University Press, 1988), 8.

34 Hansen, *Embodying Technesis,* 52.

35 Ibid.

36 Ibid.

37 Cited in ibid., 51. Further to this, Bourdieu notes that "what is 'learned by the body' is not something that one has, like knowledge that can be brandished, but something that one is" (ibid.).

38 Hansen, *Bodies in Code,* 72.

39 Ibid., ix.

40 Bernard Flynn, "Maurice Merleau-Ponty," in *The Stanford Encyclopedia of Philosophy* (2008), http://plato.stanford.edu/archives/fall2008/entries/merleau-ponty/.

41 Hansen, *Bodies in Code,* 61.

42 Ibid., 74.

43 E.g., Jennifer Gonzalez notes in "The Face and the Public: Race, Secrecy, and Digital Art Practices," *Camera Obscura* 24, no. 1 (37–65): 47–48, that Hansen turns to Agamben "to argue for digital media's potential to produce the conditions for the emergence of an identityless, subjectless singularity."

44 Indeed, this formulation echoes Derrida's reading of ethics (a term he uses under erasure), which hinges on an undecidability inherent in all decision making that nonetheless demands urgency and precipitation. As he notes in the afterword to *Limited Inc.*, 116, "a decision can only come into being in a space that exceeds the calculable program that would destroy all responsibility by transforming it into a programmable effect of determinate causes. There can be no more moral and political responsibility without this trial and this passage by way of the undecidable. [Every decision] . . . is structured by this *experience and experiment of the undecidable*."

45 Hansen, *Embodying Technesis*, 263.

46 In fact, this description of the technological real also aligns with Lacan's account of the Real as an impossible kernel. See, for example, Žižek's description of the Lacanian Real in *Sublime Object*, 169, where he describes it as (in part) a "hard, impenetrable kernel resisting symbolization" that is nonetheless linked to the embodied reality of death.

47 Hansen, *Embodying Technesis*, 263; emphasis added.

48 Indeed, this relation to deconstruction marks an important difference between Hansen and Hayles: whereas Hayles's project is, in a sense, to think past deconstruction, Hansen's writing exists prior to it, in the sense that he attempts to recoup the body as a locus of presence. In both cases, then, deconstruction stands in for a problem that needs to be solved, with both thinkers neglecting that the problem in question is precisely what deconstruction articulates. In this sense, then, we might say that both Hayles and Hansen, from opposite sides, reify deconstruction and, in so doing, fail to account for its technological dimension.

49 Lenoir, "Foreword," xxiii.

50 Barad, "Posthumanist Performativity," 809.

51 Lenoir, "Foreword," xxiv.

52 Hansen, *Bodies in Code*, 64.

53 My account of the literal material of *skulls* draws heavily from Hansen's description in *New Philosophy for New Media*, 196–205.

54 *Rapid-prototyping* is a process in which digitally produced image-models are rendered as physical objects. As the name suggests, variations of the technology are frequently used in manufacturing to quickly produce physical prototypes of computer-generated blueprints, where the latter

are often drawn in computer-aided design (CAD) programs such as QCad, ArchiCAD, and AutoCAD.

55 Robert Lazzarini, http://www.pierogi2000.com/flatfile/lazzarins.html.

56 Hansen, *New Philosophy for New Media*, 202.

57 Ibid.

58 Ibid., 199. An *anamorphosis* is a seemingly deformed image that appears normal when viewed in a particular, unconventional way. Skulls have frequently been featured in anamorphic paintings, including Hans Holbein's famous 1533 painting *The Ambassadors,* in which the distorted image of a skull in the foreground resolves when viewed from a point to the right of the painting. There is ample literature on this subject, including an excellent chapter devoted to *The Ambassadors* in Jurgis Baltrušaitis's *Anamorphic Art* (Ann Arbor: University of Michigan Press, 1977); Lacan also discusses the work in *Four Fundamental Concepts,* in which he argues that the distorted skull is the residual trace of a species of knowledge that is inaccessible to the conscious subject and that may be approached only at the boundary of the visual–imaginary order of subjectivity. I am grateful to Terry Harpold for directing me to the latter as well as for raising the tantalizing question of what anamorphosis might *sound* like, and how this sound might reveal a resonant relation (of being and language) that is more ambiguous—and thereby more "fundamental"—than Hansen's mobilization of tactility.

59 Hansen, *New Philosophy for New Media*, 209.

60 Ibid., 206.

61 Lenoir, "Foreword," xviii.

62 Hansen, *New Philosophy for New Media*, 203.

63 Ibid., 202.

64 Ibid., 203.

65 Ibid., 204.

66 Ibid., 203.

67 Ibid., 204.

68 Ibid., 211. Hansen ultimately terms this production of place the digital any-space-whatever (ASW), arguing that, as an aesthetic mediation of the digital, it describes "the priority of affectivity and embodiment in the new 'postvisual topology' of the digital age" (ibid., 205). Hansen distinguishes the digital ASW from the Deleuzian cinematic ASW as follows: "The digital any-space-whatever (ASW) is both like and unlike [Deleuze's] cinematic ASW. It is like it in that it demarcates a fundamental shift in the human experience of space, a shift from an extended, visually apprehensible space to a space that can be felt only by the body. But it

differs . . . on account of the means by which it operates this shift: whereas the cinematic ASW emerges as a transfiguration of an empirical spatial experience, the digital ASW comprises a bodily response to a stimulus that is both literally unprecedented and radically heterogeneous to the form of embodied human experience. [Simply put,] because it must be forged out of a contact with a radically inhuman realm, the digital ASW lacks an 'originary' contact with a space of human activity [e.g., the 'empty' spaces of postwar Europe]—and thus any underlying *analogy*—from which affect can be extracted" (ibid.).

69 *Sound* is unscored and site specific and has been adapted and reappropriated for various instruments and settings. In every iteration, the compositional emphasis is on achieving the "pataphysical" quality discussed later, rather than on acting a prescribed aesthetic or formal program.

70 The piece uses "extended techniques" extensively so that it does not sound like conventional piano music; however, these same techniques have been so thoroughly and frequently explored within the experimental Western art music tradition that they are by no means foreign to anyone versed in the discipline.

71 These listener accounts are (admittedly) anecdotal and were not collected via any particular methodology. However, I would argue that this shortcoming does not particularly impact the argument at hand: because *Sound* is discussed here in its conceptual—rather than aural—particularity, any reader who doubts the veracity of my account is welcome to take the discussion in the spirit of a thought experiment (i.e., "if one were to be able to stage a piece that accomplished what I claim for *Sound,* what would that mean?").

72 In "Sounding the Hyperlink," *Mosaic: A Journal for the Interdisciplinary Study of Literature* 42, no. 1 (2009): 1–18, I mount this argument more broadly, insisting that *Sound* is not a hoax in any sense, except to the extent that representation in general—and musical representation in particular— necessarily deals in deception. However, it also bears noting that, according to the American Heritage Dictionary, the word *hoax* probably comes from *hocus,* which in turn derives from *hocus-pocus,* which itself is possibly an alteration of the Latin *hoc est corpus* (this is [my] body), the words used in the Eucharist at the moment of transubstantiation. As a result, *Sound is* perhaps a hoax in the original sense of the term—meaning an alteration of material reality—but my argument still obtains with respect to the conventional use of the term (implying deception).

73 E.g., a listener who attests to hearing a low-shelf filter is saying that she hears an "unnatural" preponderance of frequencies that are higher than

a certain pitch. The observation is correct, even if the reason attributed to it is not.

74 *Pataphysics* is a term coined by the French writer Alfred Jarry that can be defined variously, including as "the science of imaginary solutions." As it is used here, the term also relates to Ted Hiebert's extrapolation of Jarry in "Nonsense Interference Patterns," in *Collision: Interarts Practice and Research*, ed. D. Cecchetto, N. Cuthbert, J. Lassonde, and D. Robinson, 103–20 (Cambridge: Cambridge Scholars Publishing, 2008), via the notion of "pata-perception," which "might be defined as 'the observation of imaginary appearances'" (121).

75 I.e., the seemingly unbounded possibilities of live electronics obsolesce illusion in the sense that they perform a claim to technical overcoming in place of the conventional sleight of hand that is implied in an acoustic setting.

76 *EQ* (short for "equalization") refers to frequency-specific amplitude modulation, allowing the listener to emphasize the treble, middle, or bass ranges of a frequency (the degree of specificity available for modulation depends on the number of bands). EQ presets, for example, are often found in commercial audio players, where they are designed to optimize the sound output for the style of music being listened to (an "Orchestra" preset might emphasize the mid- and high-range frequencies, for example, whereas an "R&B" preset might emphasize the treble and bass extremities). MaxMSP is a visual programming language that is commonly used for audio processing (though it also includes multimedia functionality) and includes EQ (as a very small and basic part of its robust capabilities). The software—which is developed and maintained by the software company Cycling '74—is designed to be highly modular, with most programs (or "patches") made by arranging and connecting "objects" (each of which has a specific function) within a two-dimensional visual canvas. An explanatory video is available from the Cycling '74 website at http://cycling74.com/products/whatismax/.

77 *Sound,* on the whole, also points to the connection—beyond the purview of this discussion, but too often ignored by sound artists—between techniques for digital manipulation and the history of music. Too often, such techniques are considered only in terms of the ways in which they modulate a signal—as a kind of ahistorical sonic manipulation—at the expense of neglecting how the modulations themselves are primed by musical histories that are as much about combining aesthetic fidelities and the materiality of specific acoustic instruments (themselves selected from other options according to their fidelity to an orchestra-based understand-

ing of music) as anything we might reasonably call "the sound itself."

78 Paul Théberge, *Any Sound You Can Imagine: Making Music/Consuming Technology* (Middletown, Conn.: Wesleyan University Press, 2007), 212.

79 Hansen, *Bodies in Code*, x–xi.

80 Hansen notes in "Media Theory," 300, that "human evolution is 'technogenesis' in the sense that humans have always evolved in recursive correlation with the evolution of technics."

81 Or rather, the body is both externalized and not; *Sound* emphasizes the latter—which is the sense in which the latter takes place—but it nonetheless remains in tension with the former (vis-à-vis the discussion of deconstruction offered earlier).

82 Importantly, this would be the case even without the computer technologies being present on stage, as the same logic applies to the physical piano as well as to the practical and conceptual architecture that binds the piece (i.e., that coalesces it as a unified object of study). In *Sound*, then, this process is simply desublimated by the represented technologies.

83 As suggested earlier and reiterated later, this perspective's departure from Hansen is subtle and occurs via the degree of strength given the deconstructive "always-already." For Hansen, this "always-already" is a soft claim that pertains to representation only rather than to the real as such.

84 Hansen, *Bodies in Code*, 176.

85 Ibid., 147.

86 This opposition between Hayles and Hansen is, of course, amplified by being described and is not intended to wholly contain the relation between the two. Instead, I am simply gesturing toward the implications of Hansen's insistence on the body's prediscursive (as opposed to extradiscursive) status as well as his tendency to figure the body as an abstract "blank slate."

87 Hansen, *Bodies in Code*, 26. Specifically, Hansen understands the body to be topologically invariant in the sense that it is a constant shared by all experience.

88 Ibid.

89 Ibid., ix–x.

90 Ibid., 26.

91 Ibid., 6.

92 Ibid.

93 Lenoir, "Foreword," xxiv.

94 Hansen, *Bodies in Code*, 15.

95 Ibid.

96 Ibid., 9.

97 I say "in the broadest possible sense" because, following Merleau-Ponty,

Hansen differentiates between the body image and the body schema, a point that is elaborated later.

98   Hansen, *Bodies in Code*, 5.

99   Ibid., 12.

100 Ibid., 9.

101 This aligns with the reading of Rafael-Lozano Hemmer's *The Trace* that I presented in chapter 4. From Hansen's perspective, we might add that the bodily reduction of telepresent works such as *The Trace* newly reveals viscerality in that the redundancy of the term (what, after all, would fail to be visceral?) is no longer supplemental (in the conventional—rather than Derridean—sense) but is now instead detached from a generalized experience of embodiment to be perceptible in its own right.

102 Hansen, *Bodies in Code*, 7.

103 Ibid., 8.

104 Ibid.

105 Ibid.

106 Mark Turner, "The Scope of Human Thought," in *On the Human* (2009), http://onthehuman.org/2004/08/the-scope-of-human-thought/.

107 Mark Johnson, "Reply to Mark Turner's 'The Scope of Human Thought,'" in *On the Human* (2009), http://onthehuman.org/2004/08/the-scope-of-human-thought/.

108 It bears noting that double-scope blending is subject to the form of critique aimed at Dawkins by Dyens: the "problem" of how we integrate alternate scales can only be framed as such through a prior disarticulation of cognition and biology. If this is the case, though, then it would imply an emergence of cultural bodies (à la Dyens) that, in turn, deconstructs the biological primacy that double-scope blending assumes.

109 Francisco Varela, "The Specious Present: A Neurophenomenology of Time Consciousness," in *Naturalizing Phenomenology: Issues in Contemporary Phenomenology and Cognitive Science*, ed. F. Varela, J. Petitot, B. Pachoud, and J.-M. Roy (Stanford, Calif.: Stanford University Press), 272.

110 Hansen, *New Philosophy for New Media*, 299.

111 Hansen, "Movement and Memory: Intuition as Virtualization in GPS Art," *Modern Language Notes* 120 (2006): 1208.

112 Hansen, *Bodies in Code*, 9. As noted previously, this movement can only be registered as such relative to the topology—itself an identity—that Hansen presumes. Here again, then, Hansen's disarticulation and opposition of terms is problematic (as discussed later).

113 Jacques Derrida, *Of Spirit: Heidegger and the Question of Technology* (Chicago: University of Chicago Press, 1989), 107–8.

114 Hansen, "Movement and Memory," 1211.

115 Ibid.

116 Wolfe, *What Is Posthumanism?*, 8.

117 Schwanitz, as cited in ibid.

118 Ibid.

119 Wolfe, ibid., xxiv, makes this point well (though in a slightly different context) when he adds to his assertion that "'First' there is noise, multiplicity, complexity, and the heterogeneity of the environment," with the caveat that he "put 'first' in quotation marks to underscore the fact that such a statement could only arise, after all, as the observation of an autopoietic system: hence 'first' here means, because of the inescapable fact of the self-reference of such an observation, 'last'; it is the environment *of* the system, not nature or any other given anteriority."

120 Hansen, *Bodies in Code,* 10.

121 Ibid., 20.

122 Hayles, "Desiring Agency," 155.

123 Hansen, *Bodies in Code,* 39.

124 Ibid., 133.

125 This insistence points to a separate line of argument, most notably in *New Philosophy for New Media,* in which Hansen takes issue with Deleuze's reading of the image in Bergson. What is at stake in that argument, as here, is Hansen's advocacy for the primacy of tactility.

126 Hansen, *Bodies in Code,* 39.

127 Gallagher, cited in ibid.

128 Francisco Varela, "Autopoiesis and a Biology of Intentionality," in *Autopoiesis and Perception Workshop* (Dublin: Dublin City University, 1992), 5.

129 Ibid., 6.

130 Hayles, *How We Became Posthuman,* chapter 10.

131 *Internet passing* consists in passing oneself off as something other than oneself online. For example, a man may pass himself off as a woman in an Internet chat room.

132 Hansen, *Bodies in Code,* 146.

133 Ibid., 157.

134 Gonzalez, "Face and the Public," 40.

135 Hansen, *Bodies in Code,* 147.

136 Ibid., 146.

137 Ibid., 147.

138 Indeed, not only is the Internet not universally accessible but also, when it is accessed, it is not done so through uniform technological constraints. That is, differences in speed, quality, content, and media constitute strata

of access that do not necessarily parallel socioeconomic striations in the flesh-world. Simply put, the materiality of the Internet has its own historical specifications so that Gonzalez, ibid., 60, summarizes Jodi Dean's argument that "while the Internet may indeed provide one site for democratic politics, it does not constitute a public sphere (particularly in the Habermasian sense of equal access and homogeneous participation)."

139 Ibid., 40.

140 Ibid., 56.

6. SKEWED REMOTE MUSICAL PERFORMANCE

1   SuperCollider is a robust open source program for audio synthesis and processing: though its functionality is similar to that of MaxMSP (discussed in chapter 5), its code-based interface sacrifices some of the latter's modularity in favor of other emphases. Limited audio documentation and a "layperson's" expanded description of SRMP can be found at http://www.davidcecchetto.net/SRMP_description.html.

2   For those who are unfamiliar with digital signal manipulation, these signal processors behave analogously to the distortion pedal that is frequently heard in popular guitar music. For example, the signal processors that were used in the premiere performance included standard digital tools such as granular processing, reversal, and pitch and time shifting. Thus a prerecorded sample of a sound that is stored in the computer as a digital file is subjected to an algorithm that alters it in a predetermined way. In the case of granular processing, for example, the algorithm divides the sample into small pieces and scatters them according to certain principles; the resulting sound contains the same sounds as the original sample but cut up and reordered in a (seemingly) chaotic fashion.

3   SRMP is a collaboration between William Brent and David Cecchetto. Networking protocols were developed by William Brent.

4   It bears emphasis that the skewing mechanism is neither ubiquitously nor uniformly operative, so the performers can no more depend on hearing a skewed representation of the remote participant's actions than they can on hearing a true representation.

5   SRMP performances have sometimes included the setup described here in conjunction with live performers, with the electronic component both articulating sounds on its own and processing those of the live instruments. In these cases, the skewing is redoubled: all the computer messages are relayed, but (because the network transfers code rather than a rendering of the acoustic signal) the processing of the local acoustic instrument is

sent to the remote location without the resulting sound. As a result, a process such as a delay applied to a local live percussionist, for example, would be applied to a remote tuba player (giving dramatically different results).

6   This claim was anecdotally demonstrated at the 2001 OpenEars festival in Kitchener, Ontario, Canada, where a recording engineer performed a test on random audience members: blindfolded participants were unable to indicate whether a sound segment had been produced by the live saxophonist sitting five feet in front of them or by the bank of loudspeakers directly behind him.

7   While these predefined spaces are affected by the live space in which they are dispersed, a qualitative difference still obtains: we might say that the two intersect rather than interact.

8   Moreover, because the specific skewing is not recorded in either rehearsals or performances, even after the fact, there is no way of knowing precisely when or in what ways it was activated.

9   Kaja Silverman, *The Acoustic Mirror: The Female Voice in Psychoanalysis and Cinema* (Bloomington: Indiana University Press, 1988), 43.

10   As I suggested in chapter 3, Hayles's mobilization of pattern and randomness offers important new language and therefore new insights into understanding the intensities at play in a digital landscape. However, although this is the case, her approach may not amount to the radical critique of deconstruction that she sometimes implies.

11   Hansen, *Bodies in Code,* 49.

12   McLuhan, *Laws of Media,* 14.

13   In this respect, McLuhan's analysis of the phonetic alphabet echoes in advance Hayles's critique of information.

14   Steven Jones, "A Sense of Space: Virtual Reality, Authenticity, and the Aural," *Critical Studies in Mass Communication* 10, no. 3 (1993): 245.

15   This is not just the case for recording technologies but also for sound synthesis. In the case of "pure" sine waves, for example, the sound would (usually) refer to a signification process (a particular way of rendering a given frequency) rather than to an original event at all.

16   In short, Serres argues in *Parasite,* 79, that since an immediate relation would not be a relation (because it would not differentiate the relata), it follows that all relation is nonrelation. As such, noise is integral to any system, including reality (where it takes the form of irrationality). See chapter 2 for a brief consideration of the "nonrelation of relation" as argued by Serres.

17   Hansen, *Bodies in Code,* 64.

18 See chapter 5 for an explanation and discussion of Hansen's claims regarding infratactility.

19 Cavell, *McLuhan in Space,* 21.

20 To be clear, in saying that sound and tactility do not differ in content, I am not neglecting the role of media in determining content but rather acknowledging that both metaphors stand in for the sum total of sensory experience (i.e., both sound and tactility are taken to be inclusive of the other senses). Moreover, it bears noting that I borrow this claim to inclusivity from McLuhan, who frequently conflates aural space and tactile experience in *Laws of Media.*

21 In this last case, the separation of video and sound is less clear, and one could make the case that principles similar to SRMP's might apply. My intention here is less to insist on a categorical difference between digital sound and video, however, than it is to show some of the ways that operational and categorical medial specificities work both with and against one another.

22 Ironically, the technology to skew animated video in real time, over distance, and without lag would likely be dependent on MIDI protocols (where MIDI is an acronym for *Musical* Instrument Digital Interface).

23 Baudrillard makes this point poetically in *The Gulf War Did Not Take Place* (Bloomington: Indiana University Press, 1995), 49, saying that "at the speed of light, you lose even your shadow."

24 In *Seduction* (New York: St. Martin's Press, 1990), 60–66, Baudrillard describes trompe l'oeil painting as "more false than the false" because it initiates a sort of "tactile fantasy" by placing the vanishing point within the viewing subject (rather than in the painting). This contrasts the privilege afforded vision in conventional painting, which creates the illusion of perspective in the object. In a meaningful sense, then, the trompe l'oeil painting "sees" the viewer, which is to say that the subject is simulated by the object. The trompe-l'oeil is more false than the false, then, because it acts as though it is representing the false appearance of the real, even as it exposes the fact that the real is nothing other than this appearance. That is, when the object "sees" the subject, for Baudrillard, the subject is revealed as a perspectival illusion.

25 Gilles Deleuze and Claire Parnet, "The Actual and the Virtual," in *Dialogues II* (New York: Continuum, 2002), 114.

26 Ibid.

27 John Shepherd and Peter Wicke, *Music and Cultural Theory* (Cambridge: Polity Press, 1997), 117. Note that there is a sense in which all resonance

is a-causal in that it depicts a continuous condition rather than a moment in time.

28 McLuhan, *Laws of Media,* 226.

29 For Hansen, this system is embodiment (broadly defined); for McLuhan, it is consciousness.

30 Cary Wolfe, *Animal Rites: American Culture, the Discourse of Species, and Posthumanist Theory* (Chicago: University of Chicago Press, 2003), 9.

CONCLUSION

1 Hayles, *How We Became Posthuman,* 4.

2 Agamben, "What Is an Apparatus?," 1.

3 Indeed, it is telling that contemporary studies such as the present one—and major texts such as Wolfe's *What Is Posthumanism?*—remain necessary more than ten years after the publication of Hayles's *How We Became Posthuman* (which I have argued is a catalyzing moment in at least the humanities strains of the discourse of technological posthumanism).

4 Agamben, "What Is an Apparatus?,"6.

5 In fact, Neil Badmington's introduction to the influential collection *Posthumanism* begins with an account of his own discomfort with the term and further cites Hassan's grudging use of this seemingly "dubious neologism" (2). Both Hayles and Dyens have similarly expressed reservations about the term itself, as a term.

6 Gilles Deleuze and Félix Guattari, *A Thousand Plateaus: Capitalism and Schizophrenia* (Minnesota: University of Minnesota Press, 1987), 7.

7 Ibid., 8.

8 Ibid., 13.

9 Ibid., 22.

10 Derrida, *Of Grammatology,* 223.

11 Maurice Blanchot, *The Writing of the Disaster* (Lincoln: University of Nebraska Press, 1986), 43.

12 Char Davies, as cited in Hansen, *Bodies in Code,* 136.

13 Ibid.

14 IBM, "Building a Smarter Planet (1 in a Series)," *The New Yorker,* February 15 and 22, 2010, 5.

15 Gilles Deleuze, *Difference and Repetition* (New York: Columbia University Press, 1995), 63–64.

INDEX

(continued from page ii)

David Cecchetto is assistant professor of critical digital theory at York University in Toronto, Canada. He is coeditor (with Nancy Cuthbert, Julie Lassonde, and Dylan Robinson) of *Collision: Interarts Practice and Research.*